GUIDEBOOK ON MACHINE LEARNING TECHNIQUES FOR ROAD QUALITY MONITORING

MARCH 2025

ASIAN DEVELOPMENT BANK

Contents

Tables and Figures

Foreword

Road quality monitoring is essential for sustainable development, as it directly affects economic growth, social equity, and environmental resilience. Well-maintained roads build crucial connections between rural and urban areas, giving access to education, health care, markets, and job opportunities, while also making communities more resilient to disasters triggered by natural hazards. Yet in many regions, particularly in rural areas of developing countries, road conditions remain poor, and road quality data are often limited, thus constraining effective infrastructure planning and investment.

The Rural Access Index (RAI) serves as a valuable metric in addressing these challenges. It evaluates the access of rural populations to reliable transportation and highlights accessibility gaps that must be bridged for sustainable and inclusive development. But a significant data availability gap limits the RAI's potential, especially in remote areas, where traditional road quality surveys can be logistically challenging and costly. Developing innovative, cost-effective methods for monitoring road quality is therefore critical to improving rural access and contributing directly to the United Nations Sustainable Development Goals (SDGs). SDG 9 in particular is centered on building resilient infrastructure, promoting inclusive and sustainable industrialization, and encouraging innovation.

To accelerate these efforts, the Japan Fund for Prosperous and Resilient Asia and the Pacific (JFPR), in cooperation with the Asian Development Bank, has emerged as a pivotal partner in driving sustainable rural development across the region. Committed to addressing urgent infrastructure and connectivity needs, the JFPR has mobilized resources and expertise to promote resilient and accessible rural areas for improved prosperity and sustainability. By supporting innovative solutions tailored to the unique needs of diverse communities, the JFPR backs initiatives that place rural areas within reach and give countless people across Asia and the Pacific a better quality of life.

With support from this fund, in collaboration with the World Data Lab, this guidebook provides a walk-through of different approaches to road quality monitoring. Outlined here is a step-by-step procedure for using geospatial data (particularly satellite imagery) and machine learning techniques to make the indicators for road quality and access more granular and timely. This procedure includes, among others, the steps outlined in the methodology used in the paper *Application of Machine Learning Algorithms on Satellite Imagery for Road Quality Monitoring: An Alternative Approach to Road Quality Surveys.* This guidebook was written by H.R. Pasindu, Aaron Thegeya, Thomas Mitterling, Clifford Njoroge, Arturo Martinez Jr., Joseph Albert Niño Bulan, Ron Lester Durante, Jayzon Mag-atas, and Oshean Lee Garonita, under the overall direction of Elaine Tan. Data and valuable technical input were supplied by the Philippine Statistics Authority and the country's Department of Public Works and Highways; Thailand's National Statistical Office, Department of Land Transport, and Department of Rural Roads; and participants in various workshops organized under this project, including the workshop on 24 August 2023 conducted in collaboration with Professor Yasuyuki Sawada from University of Tokyo. Michael Anyala, Mac Cordel, Yohan Iddawela, and Jiwoo Kang also provided insightful feedback. Manuscript editing was performed by Sean Crowley and Mary Ann Asico. Typesetting was conducted by Jonathan Yamongan. The cover was designed by Mike Cortes. Proofreading was conducted by Lawrence Casiraya. Administrative support was provided by Rose Anne Dumayas and Iva Sebastian-Samaniego.

The guidebook outlines alternatives to traditional, resource-intensive road surveys. As satellite imagery becomes more available, this methodology can help make road quality monitoring more efficient and cost-effective—particularly in resource-constrained countries.

This publication illustrates how embracing innovative technology and data-driven solutions can help accelerate progress toward a more connected and prosperous future for all.

Albert Park
Chief Economist
Asian Development Bank

Abbreviations

AI	artificial intelligence
AMD	Advanced Micro Devices
AOI	around the bounding box
API	application programming interface
ASTM	American Society for Testing Materials
CNN	convolutional neural network
CPU	central processing unit
CSV	comma-separated values
FFS	free flow speed
GB	gigabyte
GEE	Google Earth Engine
GPR	ground penetrating radar
GPS	global positioning system
GPU	graphics processing unit
Grms	root-mean-square acceleration or vehicle body acceleration
IRI	international roughness index
LiDAR	light detection and ranging
PCI	pavement condition index
PDI	pavement distress index
PSD	power spectral density

PSR   present serviceability rating

QGIS   Quantum Geographic Information System

RAI   rural accessibility index

Real-ESRGAN   Real-Enhanced Super-Resolution Generative Adversarial Network

RMS   root mean square

ROI   region of interest

RQI   ride quality index

S2   Sentinel-2

SAR   Synthetic Aperture Radar

SDG   Sustainable Development Goal

UAV   unmanned aerial vehicle

US   United States

VAE   variational autoencoder

I. Introduction

Roads are vital to the economy, serving as critical components of an area's physical infrastructure. Road infrastructures facilitate the transport of goods and services; connect rural regions to cities; and provide convenient access to educational, health care, and other key facilities. Moreover, roads foster the mobility of labor and ideas, both within and between geographic and administrative territories. The importance of access to well-built roads is reflected in the United Nations Sustainable Development Goals (SDGs) in Goal 9 on industry, innovation, and infrastructure. Target 9.1 is to *"develop high quality, reliable, sustainable, and resilient infrastructure, including regional and transborder infrastructure, to support economic development and human well-being with a focus on affordable and equitable access for all."*[1]

Quality of road pavements has two key aspects: *structural performance*, which refers to the road's structural capacity to withstand vehicular load, and *functional performance*, which refers to the comfort and skid resistance of the ride as a vehicle traverses the road. Highway engineers conduct pavement condition assessments to evaluate these aspects, providing important information about road pavement condition and actionable insights for road maintenance decisions. It is imperative for road agencies to promptly detect signs of pavement deterioration, as delays in repairs can result in rapid deterioration of affected road segments, lead to significantly higher maintenance costs, or even increase the risk of traffic accidents. Given these factors and the vital socioeconomic role roads play, it is critical to ensure that road quality monitoring and evaluation are carried out regularly.

Conventional Approaches to Evaluating Quality of Road Pavements

Several metrics may be used to evaluate quality of road pavements. These evaluations consider several aspects, including the roughness of road pavement, distress condition, structural performance, friction or skid resistance, and the presence of road markings. Depending on the aspect being measured, various types of road condition assessments may be implemented by conducting pavement condition surveys. When it comes to road roughness, the international roughness index (IRI) is a widely used metric of road quality. It is calculated using a mathematical simulation of a single car wheel moving along a given road profile at a set speed. The rural accessibility index (RAI), an indicator under SDG 9.1, measures the proportion of the rural population living within 2 kilometers of an all-season road. Some countries operationalize the concept of all-season road based on roughness. Due to its potential usefulness in providing much needed data that can facilitate enhanced monitoring of progress with respect to SDG 9.1, this guidebook will focus on the use of this particular metric in assessing road quality.

---

[1]   United Nations. 2015. *Resolution 70/1 Transforming our world: 2030 Agenda for Sustainable Development adopted by the General Assembly on 25 September 2015.* United Nations Official Document 2015.

Class I methods, which offer the highest precision in evaluating road roughness, include rod and level surveys, battery-powered profiling devices, and laser profilers. Class II methods entail the use of high-speed inertial profiler and accelerometers to evaluate roughness of the ground directly passed by a moving vehicle where the instrument is mounted. Class III methods, such as response-type systems or road meter systems, measure accumulated suspension strokes over a traveled distance, and are commonly used nowadays due to its portability and reproducibility. Finally, Class IV methods entail the use of a rating system, such as the present serviceability rating (PSR), where an engineer subjectively judges roads based on various road quality aspects (including roughness) and rates it from 5 (excellent) to 0 (very poor) and therefore tend to be less accurate than other methods. The appendix provides further details on how conventional approaches in road quality monitoring are implemented.

Challenges with Using Conventional Approaches to Evaluating Quality of Road Pavements

Conventional methods of assessing road quality involve gathering data using road survey instruments. Despite advancements in data collection technology and the expanding use of automated systems, on-site monitoring remains as the primary method of evaluating road pavement conditions. These surveys can be costly and are therefore not often conducted regularly, especially in low-income countries. This can lead to delayed detection of deterioration of road segments and higher costs for repairing the deteriorated road segments in the long run. One study estimated that neglected roads could cost up to six times more to maintain after 3 years and up to 18 times more after 5 years.[2] Moreover, road maintenance is usually underfunded, with countries spending only 20% to 50% of what is necessary. With these challenges, exploring alternative cost-effective approaches of monitoring road quality merits further investigation.

Potential of Innovative Data and Frontier Technologies

Given the resource constraints in low- and middle-income areas, this guidebook presents alternative approaches to assessing road quality by using data from satellite imagery and smartphones, and leveraging artificial intelligence (AI). The increasing availability of low-cost sensors and remote sensing technologies, as well as the proliferation of smartphone usage among users of varying technical proficiencies, provide opportunities to explore these potentially efficient and cost-effective approaches in any region in the world. Moreover, developments in AI and web technology provide a conducive environment to these innovative technologies in evaluating road quality while leveraging widely available large datasets.

This guidebook outlines the step-by-step procedure in applying a convolutional neural network (CNN) on publicly available, medium-resolution satellite imagery to predict road quality and tools that are readily accessible. The use of these publicly available analytical tools allows researchers to apply these methods for exploratory studies without incurring significant cost up-front. The guidebook also provides an overview of technical implementation of methodologies that entail the use of alternative data sources such as those from smartphones. The presentation of these innovative methods in this

---

[2]   Burningham, S. and N. Stankevich. 2005. Why Road Maintenance is Important and How to Get It Done. *Transport Notes Series*. 4. Washington, DC: World Bank. https://openknowledge.worldbank.org/handle/10986/11779.

guidebook is meant to encourage researchers and other development practitioners (particularly from the data collection arms of public works ministries and national statistics offices) to apply these techniques for exploratory studies. At this point, it is not meant to replace the conventional methods of assessing road quality which are presented in the appendix of this publication. Instead, these innovative methods supplement conventional techniques for road quality monitoring and assessment. As more development practitioners apply these innovative approaches, we could draw lessons that may help further fine-tune these new methods.

As a reminder to readers, the content of this guidebook is intended for educational use only. It is important to acknowledge that the trademarks associated with the tools and resources discussed herein belong exclusively to their respective developers, and this guidebook has no endorsement or affiliation with these companies. Readers are also advised to check the requirements associated with these resources.

II. A Review of Satellite Imagery-Based Methods in Road Quality Assessment

This section of the guidebook presents one approach for assessing road quality condition which leverages on the use of satellite imagery and AI. The proliferation of institutions with access to satellite technology has made it possible to capture images of varying resolution ranging from 10 meters to as granular as 30–50 centimeters. Moreover, developments in AI, particularly transfer learning and CNNs, allow computer vision techniques to be trained to predict road quality by detecting features from satellite imagery datasets that are correlated with varying levels of road quality.

A technical report published by the Asian Development Bank (ADB), *Exploring Machine-Learning Based Approaches for Predicting Road Quality and Compilation of Development Indicators on Accessibility,* details the findings of a feasibility study that explores the use of machine learning methods on satellite imagery and geospatial data to predict road quality and compile SDG Indicator 9.1.1 (RAI or the proportion of rural population with access to all-season roads) in select developing economies in Asia and the Pacific. The report outlines a methodology that involves the use of three types of neural networks to extract road quality features from satellite images and other preprocessed geospatial datasets presented in tabular form. Figure 1 illustrates the key steps of this methodology which are further detailed in the succeeding sections of this guidebook.

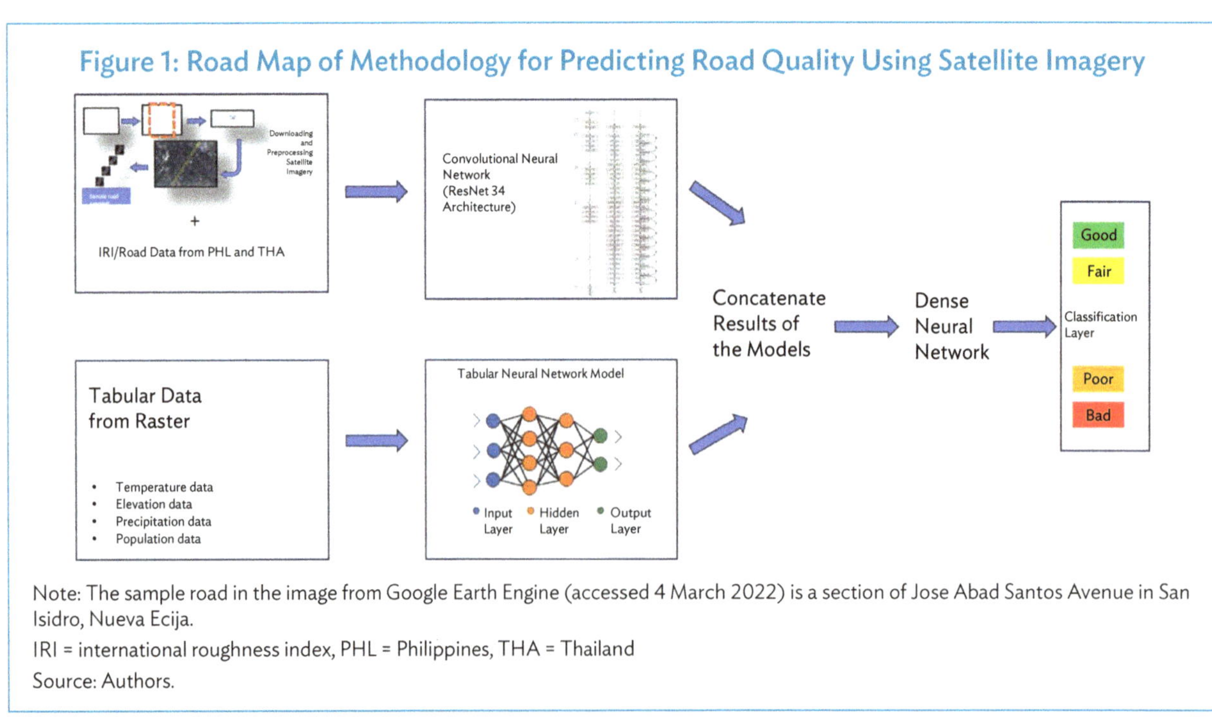

Note: The sample road in the image from Google Earth Engine (accessed 4 March 2022) is a section of Jose Abad Santos Avenue in San Isidro, Nueva Ecija.

IRI = international roughness index, PHL = Philippines, THA = Thailand

Source: Authors.

After reviewing the feasibility of adopting satellite imagery-based techniques for road pavement condition assessment, the guidebook provides a detailed walkthrough of the methodology outlined in Figure 1, by presenting step-by-step procedure of its implementation. As mentioned in the introduction, the guidebook caters to development practitioners and data scientists who are interested in applying computer vision techniques in the context of predicting a specific metric of road quality such as road roughness. It is also intended for staff of national statistics offices and transport ministries who may be responsible for compiling SDG Indicator 9.1.1 or RAI. It is important to note that such methodology is relatively new and it is not meant to replace conventional approaches for collecting road quality data but to simply complement them. Having this type of guidebook is intended to promote conduct of more feasibility studies from which lessons and insights about the methodology's potential advantages and limitations can be drawn. As an important supplement to this guidebook, users of this guidebook are encouraged to first read the abovementioned ADB technical report.

Overview of Satellite Imagery Datasets

Various types of satellite imagery datasets are widely available for download online. In the search for possible datasets that can be used for the analysis being presented in this guidebook, it cannot be avoided, particularly for nontechnical users, that many technical terminologies related to satellite data may be encountered. To provide nontechnical users with a high-level understanding on satellite datasets, an overview on the various satellite data sources and some of their technical specifications are briefly described below.

Hyperspectral and Multispectral Imagery

Hyperspectral imagery is a common type of remotely captured (or *remotely sensed*) imagery that shows information about the spectral properties of materials or objects on the earth's surface (or those above it, such as clouds). This includes the visual appearance of objects from space as seen by human eyes, which are captured in familiar photographic images of the earth's surface called *optical images* or satellite photos. But unlike in traditional photographic imaging technology which only captures the image of objects that are visible to the human eye, i.e., what is within the *visible light* spectrum, hyperspectral satellite sensors can capture data in other narrow and contiguous spectral bands that are beyond the visible light's spectral range. A good example of this is infrared imaging, which shows objects based on their emitted infrared radiation, something invisible to the naked eye but can provide information about an object's temperature and heat emission.

Each pixel in a hyperspectral image contains various information about the reflectance or emission of light at many different spectral wavelengths (thus the prefix hyper-), allowing for detailed analysis of the spectral properties of the objects in a captured scene. This data can be used to generate different images showing information about an area's vegetation, mineral deposits, water bodies, and other features based on their unique spectral signatures.

One publicly available resource for accessing hyperspectral imagery is provided by the Copernicus Program, which is headed by the European Space Agency. The Copernicus Program consists of Sentinel satellites, a constellation of satellites that take weather radar images, medium-resolution optical images, ocean and land data for environmental and climate monitoring, and air quality.

Alternatively, multispectral imagery collects data in a limited number of broader wavelength bands, typically between three and 12 bands, capturing sections like visible light (red, green, and blue), near-infrared, and sometimes shortwave infrared. This setup provides basic environmental analysis, such as vegetation monitoring and land cover classification, with relatively lower data volume and processing requirements.

A common source of medium-resolution multispectral satellite imagery is the Sentinel-2 (S2) multispectral satellite imaging mission. It has a global revisit frequency of 5 days, meaning that the same location on earth is captured at least once every 5-day period. The S2 Multispectral Instrument samples 13 spectral bands: visible and near-infrared at a resolution of up to 10 meters, red-edge and short-wave infrared at a resolution of 20 meters, and atmospheric at a resolution of 60 meters. Conversely, Landsat satellite images of the earth's surface are at a 30-meter resolution with a revisit frequency of about once every 2 weeks, including multispectral and thermal data. The most recent Landsat data are in Collection 2, which has two catalogues, 7 and 8. Landsat 8 has improved calibration, signal to noise characteristics, radiometric resolution, and spectrally narrower wavebands than Landsat 7.

It should be noted here that in the context of satellite imagery datasets, the term spatial resolution generally refers to the ground area that a single pixel in a satellite or aerial image represents. If a satellite image has a spatial resolution of 10 meters, each pixel in the image represents a 10x10 meter area on the ground. A rule of thumb is that higher resolution is generally better for analysis. It is instructive to note that from a human's visual perspective, individual defects within a road, such as surface cracks or potholes, are not visible to the naked eye in both the publicly available and high-resolution satellite data. However, high-resolution satellite image data allows for the identification of the road type.[3]

Synthetic Aperture Radar Imagery

Aside from hyperspectral imagery, another remotely sensed data that can be used is imagery captured from Synthetic Aperture Radar (SAR). Unlike hyperspectral imagery which shows captured images of the earth by orbiting satellites through reflected energy coming from the sun, SAR imagery is produced using an active energy-emitting sensor aimed toward the earth's surface, where the energy reflected upon interaction with the earth's surface is captured into the imagery. SAR can also capture specialized images using longer wavelengths in the electromagnetic spectrum to be able to characterize specific attributes of the earth's surface, such as vegetation cover, surface water, human-made structures, as well as soil and ice. Its use of longer wavelengths also allows it to remotely capture data even through cloud cover. SAR's versatility in capturing different aspects of the earth's surface beyond what optical images can provide makes it useful in many environmental applications, such as in biogeography, geology, environmental engineering, and archeology, to name a few. For the proposed analysis approach presented here, SAR imagery may be used as a source dataset for deriving elevation and slope data, which can then supplement the visual road roughness information coming from the hyperspectral imagery.

Similar with hyperspectral imagery, one good example of a source for global SAR imagery is the European Space Agency, particularly the Sentinel-1a satellite, which has provided free SAR data to the public since its launch in 2014.

---

3   C. Sabottke and B. Spieler. 2020. The Effect of Image Resolution on Deep Learning in Radiography. *Radiology: Artificial Intelligence*. 2 (1). https://doi.org/10.1148/ryai.2019190015.

A special type of SAR imagery called Interferometric Synthetic Aperture Radar (InSAR) is useful for monitoring and mapping ground deformation, which can be used to detect issues with road quality. Providing high precision, InSAR analyzes the temporal phase difference between radar signals captured from multiple satellites that pass over the same location. When these signals are compared, InSAR can detect even minute shifts in the earth's surface over time, which makes InSAR valuable for detecting ground displacement caused by land deformities such as landslides and ground subsidence.

Other Geospatial Datasets

As mentioned above, there are other imagery datasets derived from hyperspectral or SAR imagery which may be used for the imagery-based analysis of road quality assessment. These additional datasets supplement the optical imagery data derived from the hyperspectral imagery as they are likely to influence road quality. These datasets include the following:

- *Average daily temperature, e.g., may be derived from the European Centre for Medium-Range Weather Forecasts Reanalysis Version 5 (ECMWF-ERA5), captured at a height of 2 meters with a resolution of 0.25 arc degrees.*
- *Average total precipitation, e.g., may be derived also from the ECMWF-ERA5, with a resolution of 0.25 arc degrees.*
- *Digital Elevation Model, e.g., may be derived from National Aeronautics and Space Administration's (NASA) Shuttle Topographic Mission (SRTM), with a resolution of 30 meters.*
- *Total population, e.g., available from Worldpop, estimated per 100-square-meter grid.*

Temperature and precipitation are important factors that influence how quickly a road degrades. In addition, land gradient determines land surface flow of water, which over time impacts road quality. Finally, the population within the neighborhood of a road segment is an indicator of the frequency of use of a given road segment.

Similar Applications

A few applications yet insightful initiatives featuring use of satellite imagery for monitoring road quality have previously been demonstrated using the similar approach showcased in this text. Oshri et al.,[4] for instance, trained a CNN to binarily classify road quality using satellite imagery from Landsat 8, Sentinel 1, and actual gathered road quality data from the Afrobarometer Round 6 survey, achieving a 70.5% accuracy on road quality predictions. Another notable study, which is the primary reference for the algorithm built in this text, is by Cadamuro, Muhebwa, and Taneja (2018).[5] They also leveraged CNN trained on satellite images and ground-collected road roughness data for them to be able to predict the quality of roads in Kenya using a binary, and a five-category classification. They were able to train the models that achieved a prediction accuracy of as much as 90% for the binary classification, and 73% for the five-category classification.

[4]   B. Oshri et al. 2018. *Infrastructure Quality Assessment in Africa using Satellite Imagery* and Deep Learning. CoRR.

[5]   G. Cadamuro, A. Muhebwa, and J. Taneja. 2018. Assigning a Grade: *Accurate Measurement of Road Quality Using Satellite Imagery*. arXiv. https://doi.org/10.48550/arXiv.1812.01699.

Other related studies using CNN with other imagery-based datasets for monitoring road quality may also be noted. In particular, Leduc and Assaf[6] also used CNN and automated machine learning (AutoML) models to identify and classify road deformations based on road images captured using a GoPro camera, achieving an average of 71% in identifying true positives. In addition, Eisenbach et al.[7] also predicted the presence of road distress level using CNNs trained on high-definition road surface images taken from German highways.

Feasibility of Remote Sensing Techniques Using Satellite Imagery

Key Advantages

Using satellite imagery as a preliminary tool for assessing road quality conditions may be cost-effective and efficient in settings wherein immediately conducting on-site road survey methods is not feasible. If implemented properly, satellite imagery-based methods may facilitate more regular road quality assessments as satellite images are collected frequently.

Satellite imagery-based road quality assessment is also less labor-intensive and therefore can be implemented in countries with low number of skilled personnel in road quality monitoring. Furthermore, with the regular availability of updated satellite imagery over areas covered by roads, this approach allows for a more frequent and dynamic measurement of road quality over time. In terms of precision, results from testing this methodology prove to be promising as well in that it allows the assessment of road segments at a level of detail of up to 100 meters even when using publicly sourced satellite data.

Limitations

A common result among the studies previously cited above is the inability to identify the type of pavement surface distress or settlement problems without further on-site investigation. This presents a major limitation for the imagery-based road quality assessment being described here. Certain technical limitations in the datasets to be used must also be taken into consideration in undertaking the satellite imagery-based road quality assessment. In the use of SAR imagery for instance, certain properties specific to SAR such as its side-looking nature, intrinsic constraints such as foreshortening, with geometrical decorrelation caused by changes in viewing angles and temporal decorrelation due to changes in surface properties over time along with additional atmospheric effects, speckle noise which tends to reduce the image resolution and contrast, and changes in backscattered signal intensity, may lead to complications in its use and interpretation compared to that derived from hyperspectral imagery. Conversely, in the case of hyperspectral imagery, certain data-related issues such as lower image resolutions for certain sources, spectral mixing, presence of imagery artifacts, may also be present in the data since one pixel of a satellite camera may represent more than five meters of different materials or objects on earth. The quality of the actual road condition data, which is considered as the ground truth data for training and validating the CNN models should also be

---

6   E. Leduc and G. Assaf. 2022. Road Visualization for Smart City: Solution Review with Road Quality Qualification. *Internet of Things. 12 (100305)*. https://doi.org/10.1016/j.iot.2020.100305.

7   EM. Eisenbach et al. 2019. *Enhancing the Quality of Visual Road Condition Assessment by Deep Learning*. 26th World Road Congress. Abu Dhabi, UAE.

managed, especially if a subjective assessment or other condition evaluation techniques of lower accuracy levels are used by the source agency. In addition, the actual road condition data which is sourced from different agencies highly vary in terms of road segment lengths with a uniformly recorded roughness index (or any other similar pavement condition metric).

Because of these potential issues, proper handling and preprocessing of the imagery and the road quality data must be guaranteed from the source or performed correctly (in the case of raw datasets) prior to performing the actual analysis.

Kindly note that the potential issues are only associated with the use of free multispectral data. Purchasing high-resolution satellite imagery is still an option, however, this may require significant monetary resources.

Applicability

Satellite-based monitoring efforts will be at best, complementary to current monitoring efforts and will not replace on-site condition monitoring.

There is potential for satellite imagery to be used for preliminary screening of the roads which are in good condition. The evidence available at present on using satellite imagery for road condition monitoring proves that the roads can be categorized sufficiently accurately if the number of classes are low (around three, e.g., good, satisfactory, bad) and those roads that are classified as 'good' have a reasonable level of precision. This would suggest that satellite-based monitoring may reduce unnecessary vehicle-based site inspection trips and that the monitoring effort providing continuous satellite-based imagery will allow prioritization for the timely detection of pavement and infrastructure problems. This would be possible with medium- or high-resolution satellite images.

There is limited evidence on the use of satellite images to replace condition monitoring metrics such as pavement roughness, and pavement distresses for a large-scale network. This would require the use of several high-resolution satellite images and a computationally extensive process that may not be practical for a large network.

The image processing techniques to analyze satellite images to evaluate road conditions are also well established in the several research studies that have carried out, however the accuracy remains an issue, especially when there are high number of road condition categories.

Requirements

The main requirement is to determine the type of satellite imagery that is required for the condition monitoring. As discussed in the preceding chapter, if simple categorization of the road condition is the requirement, the use of freely available satellite image sources (medium-resolution) can be utilized. However, the date of image capture and other disturbances, such as cloud cover, need to be considered. The use of SAR imagery and information available from other spectral bands can be considered as well for this purpose. However, opting for free image sources would require more time for preprocessing work.

Therefore, it is recommended to acquire high-resolution imagery that is used for road network condition monitoring. This will guarantee that computational errors are minimized and specific areas covering the road network can be acquired to reduce the cost of purchasing. Moreover, with high-resolution images, manual verification would be even possible if there are sections where results are inconclusive. The image acquisition cost could be reduced if the images are shared between other state agencies such as irrigation, agriculture, and forestry, which may need the information from the satellite images.

The cost of a high-resolution image is approximately $20 per square kilometer (km^2).[8] However, this depends on the area of the image purchased. The cost varies from approximately $40 per km^2 to $11 per km^2 when the land area increases from 6.5 km^2 to 193 km^2. For example, an area spanning 2,000 km^2 may already cost roughly $20,000 to procure a high-resolution image, an amount that may not be readily available to government researchers who are interested to pilot-test satellite imagery-based methods for exploratory studies.

The other important aspect is to identify how the computation component is carried out. The road agency must determine whether it will be done in-house or contracted to a specialist firm. This is similar to the existing pavement condition monitoring regimes which are either carried out by in-house staff or through consultancy companies. However, the capacity building of internal staff involved in the road asset management division must be done to provide the required knowledge.

Prior to implementation, the machine learning model should be validated for local conditions since it can vary considerably for different road types, terrain conditions, land use, and weather conditions. This may be necessary at the road agency level so that the model can be trained and validated for the road network maintained by the particular road agency.

The computational resources for image processing are explained based on the following example:

The imagery datasets consist of the raw (spectral unmixing not applied) S2 images (S), unmixed S2 images (SU), combined Sentinel-2 images (SC) (unmixed stacked together with the raw ones), and Planet images (P). Total of 795 and 619 patched images for Planet and Sentinel-2, respectively, were used in the machine learning model dataset (training, validation, and testing). All the experiments were conducted using a 12-core Intel® Core i7-9750H central processing unit (CPU) @ 2.6 gigahertz (GHz) with 16 gigabytes (GB) RAM and a single NVIDIA Quadro T1000 graphics processing unit (GPU) with 4 GB dedicated memory. The U-Net model is accelerated by the GPU (and partially by the CPU for processing data feed), while the CPU undertakes the training of the random forest model.[9]

Therefore, the computation resources required are not a major issue and should be available in most road agencies if the analysis is to be done in-house.

---

[8]   Apollo Mapping. nd. *Worldwide-4 Satellite*. https://apollomapping.com/worldview-4-satellite-imagery.
UP42. 2023. *A definitive guide to buying and using satellite imagery*. https://up42.com/blog/tech/a-definitive-guide-to-buying-and-using-satellite-imagery.
American Association for the Advancement of Science. nd. *High-Resolution Satellite Imagery Ordering and Analysis Handbook*. https://www.aaas.org/resources/high-resolution-satellite-imagery-ordering-and-analysis-handbook.
[9]   B. A. Gebreegziabher. 2021. *Mapping Road Pavement Quality from Optical Satellite Imagery using Machine Learning*. Enschede, The Netherlands: Faculty of Geo-Information Science and Earth Observation of the University of Twente.

III. Application of Machine Learning Algorithms on Satellite Imagery for Road Quality Monitoring

The next section of this guidebook is targeted for researchers and other development practitioners (particularly from national statistics offices and transport ministries) who wish to examine the feasibility of using machine learning algorithms to supplement conventional road quality monitoring methods. The steps presented here assume that publicly available satellite imagery (whose resolution is lower than proprietarily sourced satellite images) are used in the reader's exploratory study; hence, there is a section on super-resolution.

The objective of super-resolution is to enhance the spatial resolution of images originally captured at a lower quality, specifically increasing the resolution from 10 meters per pixel to a much higher quality ranging from 2.5 meters to 5 meters per pixel. This enhancement process is crucial as it is aimed at improving the accuracy of image classification.

To achieve this resolution enhancement, we leverage a sophisticated training model known as Real-Enhanced Super-Resolution Generative Adversarial Network (Real-ESRGAN). Real-ESRGAN is a deep-learning model designed to handle super-resolution tasks, allowing it to learn from data and generate high-resolution images from low-resolution inputs. This model's capabilities are pivotal for the task at hand.

For our dataset, we curate a paired collection of images from the National Agriculture Imagery Program and combine them with United States (US) TIGER/Line Road Census shapefiles and S2 satellite imagery. This dataset, featuring corresponding pairs of images, serves as the input for the Real-ESRGAN model. The purpose is to train the model to perform super-resolution, essentially enhancing the imagery by a factor of four.

Super-resolved images serve as inputs for image classification. We approach this in two distinct phases. The first phase is a comprehensive four-class classification task aligning with IRI ratings, which include categories such as 'bad', 'poor', 'fair', and 'good.' The second phase simplifies this into a two-class classification task, merging the categories 'bad' and 'good.' Image classification is a critical component of our work, as it involves categorizing the enhanced images based on their quality or condition, thereby allowing the use of the model to predict road quality in areas without survey data.

However, the analysis reveals no significant difference in classification accuracy between low-resolution and super-resolved images, even though the latter exhibits visibly enhanced quality. This outcome suggests that, while super-resolution techniques improve visual clarity, they may not necessarily contribute to higher classification performance in this context. Nonetheless, sharing this methodology (see section VI) may still be valuable for the research community, as it provides insight into the effectiveness of super-resolution for classification tasks. Future studies may benefit from this data, potentially leading to improved techniques or applications that may yield different results. Additionally, future studies that apply super-resolution in contexts different from the case study presented here may yield more satisfactory results. This potential makes this guidebook a valuable educational reference.

Users are also advised to check for updates to the software and services referred to and pictured in screenshots in this guidebook. The discussions in this guidebook are meant for educational purposes. It should be noted that trademarks of tools and resources used are owned solely by the respective developers, and this guidebook is not endorsed by or affiliated with these companies in any way.

Neural Networks

There has been an increasing application of neural network technology for road quality detection, due to an increase in computational power and memory, and the collection of large datasets. The key networks used include CNNs and long short-term memory recurrent neural networks (LSTM-RNNs). LSTMs are particularly useful for identifying patterns in sequential data.

Since 2012, there has been an exponential improvement in the prediction capacity of neural networks by using deeper architectures that take advantage of large datasets, leading to their ubiquitous use in many applications.

Neural networks, often referred to as artificial neural networks (ANNs) or simply neural networks, are a fundamental class of machine learning models inspired by the structure and functioning of the human brain. They are composed of interconnected nodes, or artificial neurons, organized into layers. Neural networks are used for a wide range of tasks, including regression, classification, pattern recognition, and function approximation.

The basic components of a neural network include:

1. **Neurons** (Nodes): Neurons are the basic computational units of a neural network. They receive input signals, perform a weighted sum of these inputs, and apply an activation function to produce an output signal.

2. **Layers:** Neurons are organized into layers. The most common layers in a neural network include:
   - Input Layer: Receives input data and passes it to the hidden layers.
   - Hidden Layers: Intermediate layers that perform computations on the input data. These layers are responsible for learning and extracting features from the data.
   - Output Layer: Provides the final output of the network. The number of neurons in the output layer depends on the specific task, such as binary classification (1 neuron) or multi-class classification (multiple neurons).

3. **Weights and biases:** Each connection between neurons has a weight associated with it. The weights determine the strength of the connections between neurons. Additionally, each neuron typically has a bias term that helps in adjusting the output of the neuron.

4. **Activation functions:** Activation functions introduce non-linearity into the network, allowing it to model complex relationships. Common activation functions include the sigmoid, Rectified Linear Unit (ReLU), and hyperbolic tangent (tanh).

Generative Models

Generative models include a variety of models and strategies aimed at generating fresh data or content that resembles human-created material. There are various classes of generative AI models, each with its own distinct approach to content generation. It is important to note that these models can "hallucinate" or produce inaccurate data, as the outputs are estimates generated based on the data they were trained on. Varshney[10] defined these hallucinations as a generation of texts or responses that seem sound, fluent, and natural, but are factually incorrect, nonsensical, or unfaithful to the provided source input. Therefore, proper testing and validation are essential to ensure the reliability and accuracy of the results. These include:

Variational Autoencoders (VAEs)

VAEs are generative models that learn to encode data into a latent space before decoding it to recreate the original data. In this context, the "latent space" is like a hidden or secret space where the VAE learns to store and represent important information from the data. This space captures the essential features and patterns of the data in a more compact form, making it easier for the VAE to generate similar data when needed. They produce new samples from the learned distribution by learning probabilistic representations of the input data. VAEs are frequently utilized in image-generating jobs, but they have also been used in text and audio generation.

Generative Adversarial Networks (GANs)

GANs were introduced by Ian Goodfellow and other researchers at the University of Montreal. They are algorithmic architectures that use two neural networks, pitting one against the other in order to generate new, synthetic instances of data that can pass for real data.

Real-ESRGAN

Real-ESRGAN Is an advanced deep-learning neural network used for image super-resolution, enhancing the quality and details of low-resolution images.

---

10 B. A. Gebreegziabher. 2021. Mapping Road Pavement Quality from Optical Satellite Imagery using Machine Learning. Faculty of Geo-Information Science and Earth Observation of the University of Twente.

Figure 2: Comparison Between ESRGAN and Low-Resolution Images

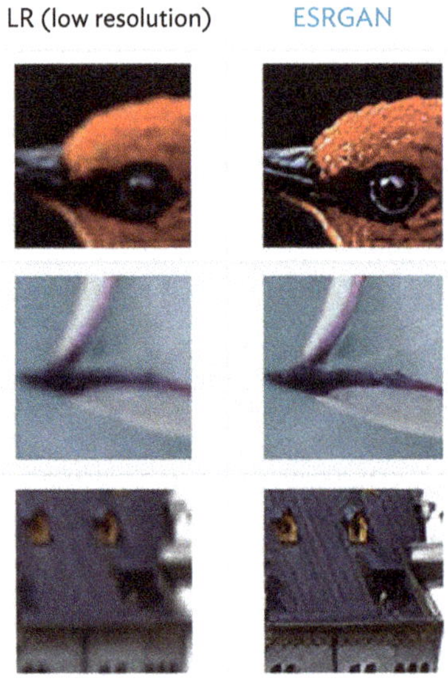

ESRGAN = Enhanced Super-Resolution Generative Adversarial Network.
Source: X. Wang, et al. 2018. ESRGAN: Enhanced Super-Resolution Generative Adversarial Networks.
https://arxiv.org/abs/1809.00219.

Convolutional Neural Networks

CNNs or ConvNets are a class of deep-learning models designed for processing and analyzing structured grid data, such as images and video. They are particularly well suited for tasks involving visual recognition, image classification, object detection, and image generation. CNNs have been pivotal in advancing the field of computer vision.

The key features of CNNs include:

(i) **Convolutional layers:** These layers use convolutional operations to scan input data with learnable filters (also known as kernels or feature detectors). Convolution is a mathematical operation that extracts local patterns or features from the input data, which makes CNNs excellent at capturing spatial hierarchies. Convolutional layers help identify edges, textures, and other visual patterns in images.

(ii) **Pooling layers:** Pooling layers reduce the spatial dimensions of the feature maps produced by convolutional layers. Common pooling operations include max pooling and average pooling, which help reduce the computational complexity and provide some degree of translational invariance.

(iii) **Fully Connected layers:** After several convolutional and pooling layers, CNNs often employ fully connected layers (also known as dense layers) to make predictions or classifications. These layers can learn complex, nonlinear relationships in the extracted features and produce the final output.

(iv) **Activation functions:** Nonlinear activation functions, such as ReLU, are applied after convolutional and fully connected layers to introduce nonlinearity into the model and enable it to learn complex patterns.

Types of CNNs used in the study include:

Residual Networks

Residual networks are a type of deep-learning model that uses skip connections to overcome the problem of vanishing or exploding gradients.[11] Skip connections are connections that bypass some layers and add the input of a layer to the output of another layer. This way, the network can learn residual functions that are easier to optimize and can avoid degradation of performance when the network becomes deeper. Residual networks were proposed by researchers at Microsoft Research in 2015 and won the ImageNet competition that year.[12]

Efficient Networks

EfficientNets are a type of CNN that uses a novel scaling method to achieve better accuracy and efficiency. The scaling method uniformly scales the depth, width, and resolution of the network using a compound coefficient that balances the trade-off between performance and resources. EfficientNets have been shown to outperform other CNNs on various image classification tasks, while being much smaller and faster.[13]

---

[11] Vanishing or exploding gradients are issues that occur during the training of deep neural networks, where gradients become exceedingly small in the early layers or grow excessively large, respectively, as they are propagated backward through the network. As gradients shrink closer to zero in the early layers, the weights in those layers receive very minimal updates, making it difficult for the network to learn. If gradients become very large, they cause drastic updates to the network weights, which can lead to instability in training, poor convergence, or the model's parameters "blowing up" to very high values. Both will result to a poorly trained model that have degraded or unpredictable performance.

[12] H. Kaiming et al. 2015. *Deep Residual Learning for Image Recognition.* arXiv. https://doi.org/10.48550/arXiv.1512.03385.

[13] M. Tan, R Pang, and Q. Le. *EfficientDet: Scalable and Efficient Object Detection.* Proceedings of the IEEE/CVF Conference on Computer Vision and Pattern Recognition (CVPR), 2020, 10781–10790.

IV. Hardware and Software Requirements and Setup

Hardware

The minimum system requirements are:

2.5 gigahertz 4-core processor or better, 16 GB RAM, 200 GB of free hard disk space, CUDA/ROCm capable GPU with Tensor-cores/tensorfloat32 capabilities if doing local training, a reliable internet connection (minimum upload speed 10 megabytes (MB), minimum download speed 10 MB).

Software

- Python 3.8+ (local resource)
- ArcGIS/Quantum Geographic Information System (QGIS) (local resource)
- Earth Engine API (local/ cloud)
- Geemap (local/cloud)
- PyTorch (local/cloud)
- Basic_SR (local/cloud)
- Git (local/cloud)
- Google Drive with at least 100 GB space, Google Colab, and a Gmail account (cloud)
- Google Chrome browser version 79.0.3945 or higher

Software Requirements and Installation

Key geospatial objects include vector and raster files. Vector data represents geographic data symbolized as points, lines, or polygons. Raster data represent geographic data as a matrix of cells that each contains an attribute value. While the area of different polygon shapes in a dataset can differ, each cell in a raster dataset is the same cell.

Use of geospatial data requires you to have some GIS software installed on your machine. In this section, we provided detail instructions on installing the use of QGIS. You also have the option to use ArcGIS. However, it should be noted that ArcGIS is proprietary software, while QGIS is open source.

Kindly note that some software applications need a license depending on the type and size of the organization. Please ensure to use licensed software applications only.

QGIS

To install QGIS follow the steps below:

1. Go to the QGIS download page (https://www.qgis.org/en/site/forusers/download.html).
2. You will see two versions, the "Standalone Installer"[14]
     and the "OSGeo4W Network Installer".[15] If you're not sure which one to choose, you can usually go with the "Standalone Installer" for most users. It is recommended to choose the Long-Term Release version of QGIS as it has less bugs than the latest version.

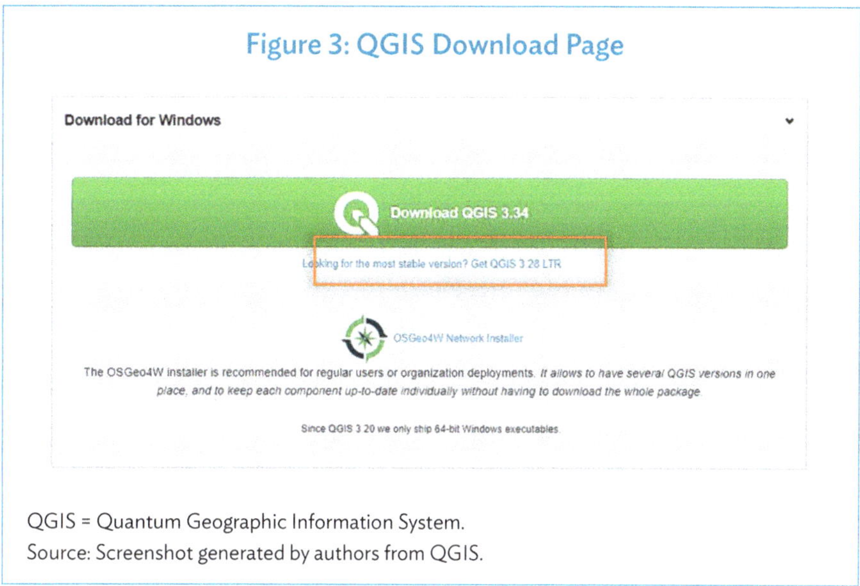

Figure 3: QGIS Download Page

QGIS = Quantum Geographic Information System.
Source: Screenshot generated by authors from QGIS.

3. Click on the "Standalone Installer" link.
4. Download the installer for your version of Windows (32-bit or 64-bit, usually laptops run at 64-bit but it is best to check your machine specifications). To check your system type, you can right-click on "This PC" or "My Computer," select "Properties," and look for the system type information.
5. Run the downloaded installer.
6. Follow the installation wizard, and you can generally accept the default settings. You can customize the installation location if needed.
7. Once the installation is complete, you can launch QGIS from your Start menu.

---

[14] Using "Standalone Installer" is the easiest way to install QGIS which is recommended for beginners.
[15] "OSGeo4W Network Installer" make it possible to install several versions in parallel and to do much more efficient updates as only changed components are downloaded and installed. This is recommended for more advanced QGIS users.

Python

Python is used to build Python packages. It allows the user to create a necessary programming environment with which one can develop applications.

The step-by-step procedure to download and install Python is as follows:

Installing Python Packages

You can install individual Python packages using the steps detailed below. This is done by invoking the following commands within the command prompt or Jupyter kernel:

→   pip install <your package>

Note: On Linux/Mac OS, make sure to denote your Python version before pip or python, i.e., 'pip3' or 'python3.'

AI Libraries

Effective processing of big data and the use of deep-learning AI models frequently require additional computing resources. Processing times can be significantly reduced by employing the use of GPUs. In this section, we detail parallel computing platforms to use GPU as well as AI libraries within Python.

CUDA

Developed by NVIDIA, CUDA is a parallel computing platform and application programming interface (API) that allows developers to harness the computational power of NVIDIA GPUs. It provides a set of libraries, APIs, and tools to enable GPU-accelerated computing. It is widely used in various domains, including scientific simulations, deep learning, machine learning, and scientific computing.

CUDA is highly compatible with NVIDIA GPUs and is considered one of the most popular and well-supported GPU programming frameworks. Developers can write CUDA code using the CUDA C/C++ or CUDA Python language bindings.

Key libraries and tools associated with CUDA include:

- cuBLAS (for linear algebra)
- cuDNN (for deep learning)
- cuFFT (for Fast Fourier Transuform).

ROCm

It is an open-source, Heterogeneous System Architecture-compliant platform for AMD GPUs, developed by AMD. It aims to provide a unified platform for programming AMD GPUs for various computing workloads.

ROCm supports a range of programming languages and APIs, including C/C++, Heterogeneous-compute Interface for Portability, and OpenCL. It is designed to be more open and vendor-agnostic, allowing for broader compatibility with different GPU architectures and hardware.

Key libraries and tools associated with ROCm include:

- ROCm Runtime
- ROCm Compiler
- ROCm Math Libraries
- ROCm Developer Tools

ROCm has gained popularity in scientific and high-performance computing, as well as in machine learning and deep learning, as an alternative to CUDA for those using AMD GPUs.

PyTorch

PyTorch is an open-source deep-learning framework developed by Facebook's AI Research lab (FAIR). It is known for its flexibility, dynamic computation graph, and ease of use, which has made it one of the most popular libraries for deep learning and machine learning tasks.

Key features and characteristics of PyTorch include:

- Dynamic computational graph: PyTorch uses a dynamic computation graph, meaning that the graph is constructed on the fly as operations are performed. This allows for more flexibility in defining and modifying the network architecture during runtime, making it well suited for tasks like natural language processing and reinforcement learning.
- Tensor computation: PyTorch is built on the concept of tensors, which are multi dimensional arrays that are similar to NumPy arrays. Tensors are fundamental to deep learning, and PyTorch provides a variety of tensor operations for efficient mathematical computations.
- Neural network library: PyTorch includes a neural network library that simplifies the process of building and training deep neural networks. It offers a wide range of predefined layers, activation functions, loss functions, and optimization algorithms.
- Autograd: PyTorch provides automatic differentiation through its autograd package. This means that it can automatically calculate gradients, which are essential for training neural networks using backpropagation. This feature simplifies the implementation of custom loss functions and network architectures.

Here are the simplified steps to install PyTorch built with CUDA:

1. Install CUDA Toolkit:
   - Visit the CUDA Toolkit website: https://developer.nvidia.com/cuda-downloads.
   - Download the CUDA Toolkit installer for Windows.

Code on Checking CUDA Toolkit

```
(base) PS C:\Users\cliffordkleinsr> nvcc --version
nvcc: NVIDIA (R) Cuda compiler driver
Copyright (c) 2005-2022 NVIDIA Corporation
Built on Wed_Sep_21_10:41:10_Pacific_Daylight_Time_2022
Cuda compilation tools, release 11.8, V11.8.89
Build cuda_11.8.r11.8/compiler.31833905_0
(base) PS C:\Users\cliffordkleinsr> |
```

Figure 4: CUDA Toolkit

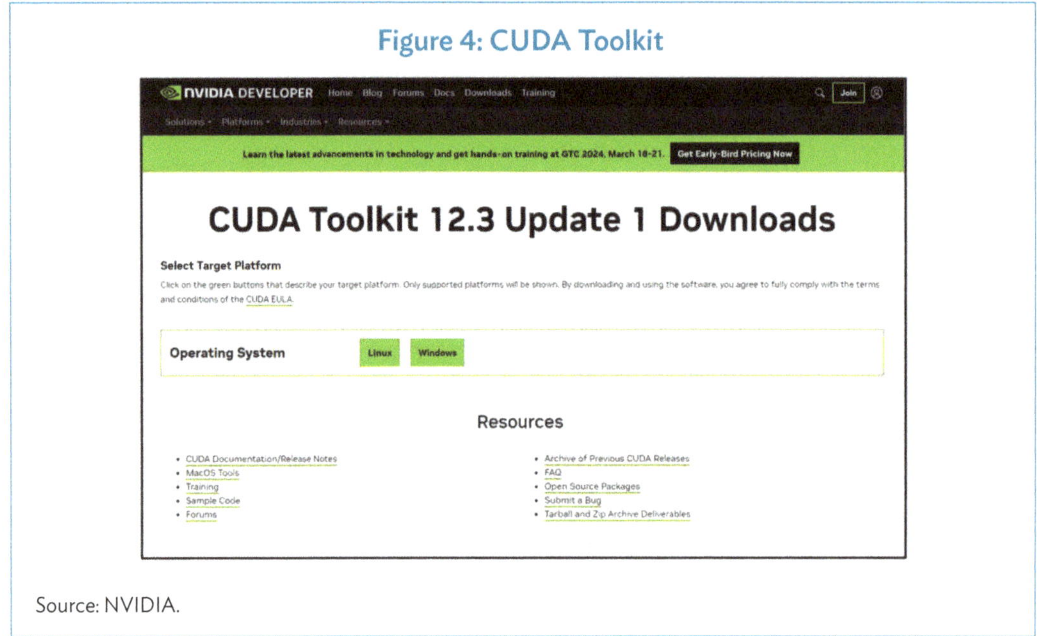

Source: NVIDIA.

- Run the installer and follow the prompts, selecting the desired options for installation.

2.   Install cuDNN:
   - Go to the NVIDIA cuDNN website (https://developer.nvidia.com/rdp/cudnn-download).
   - Download the cuDNN version compatible with your installed CUDA Toolkit (you may need to create an NVIDIA Developer account if prompted).
   - Extract the downloaded cuDNN package.
   - Copy the extracted files (`bin`, `include`, `lib`) into the corresponding CUDA Toolkit installation directories.

3.   Install PyTorch and related libraries:
   a. To install the PyTorch framework use the available mirrors provided on the Pytorch official website (https://pytorch.org/):

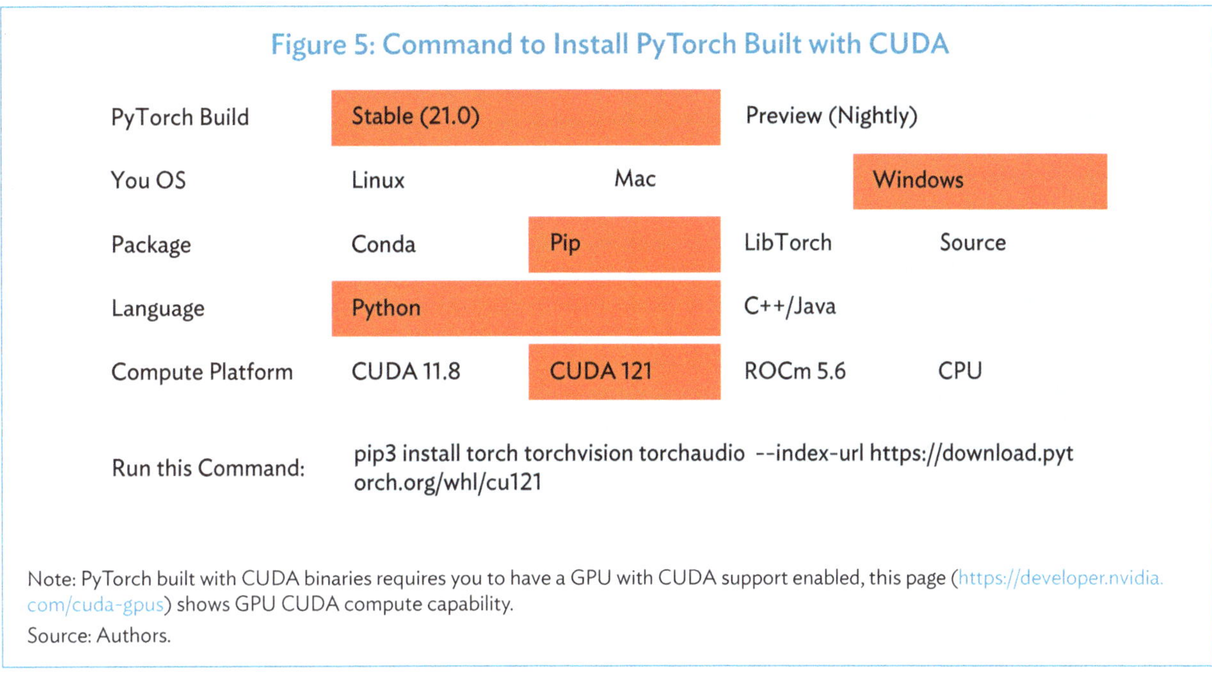

Note: PyTorch built with CUDA binaries requires you to have a GPU with CUDA support enabled, this page (https://developer.nvidia.com/cuda-gpus) shows GPU CUDA compute capability.
Source: Authors.

If you are running an AMD GPU device, you should install PyTorch with ROCm build support (https://docs.amd.com/bundle/ROCm-Deep-Learning-Guide-v5.3/page/Frameworks_Installation.html). To confirm PyTorch is built with CUDA/ROCm binaries, run the following on your shell:

→ python

→ import torch

→ torch.cuda.is_available

Code on Confirming Pytorch is Built with CUDA/ROCm Binaries

```
(base) C:\Users\Peter>python
Python 3.9.12 (main, Apr  4 2022, 05:22:27) [MSC v.1916 64 bit (AMD64)] :: Anaconda, Inc. on win32
Type "help", "copyright", "credits" or "license" for more information.
>>> import torch
>>> torch.cuda.is_available()
```

If the command returns **True**, then you have successfully installed PyTorch with CUDA/ROCm capabilities.

Chrome Browser

To install Google Chrome Web Browser version 79.0.3945 or higher, follow the step-by-step procedure below:

1. Open your preferred web browser on your computer.
2. Go to the official Google Chrome website. You can search for "Google Chrome" or directly access the website (https://www.google.com/chrome/).
3. On the Google Chrome website, you should see a "Download Chrome" button. Click on it.
4. The website will detect your operating system automatically and offer you the appropriate version of Chrome to download. If you want to download Chrome for a different operating system, you can use the "Other Platforms" link to select your desired version.
5. Once you click on the "Download Chrome" button, a Terms of Service window may appear. Read through the terms, and if you agree, click the "Accept and Install" button.
6. The Chrome installer file will be downloaded to your computer. Locate the downloaded file (usually found in the Downloads folder).
7. Double-click on the installer file to run it. This will start the installation process.
8. A User Account Control (UAC) prompt may appear, asking permission to change your computer. Click "Yes" to continue.
9. The installer will begin the installation process, which may take a few minutes. You may see a progress bar indicating the installation progress.
10. Once the installation is complete, Chrome will open automatically, and you will see a welcome page.
11. You can choose to sign in to Chrome with your Google Account to sync your bookmarks, history, and other settings across devices. Alternatively, you can skip this step by clicking "Skip" or "No Thanks."

Google Account

Setting up a Google Account involves the following step-by-step procedure:

1. Go to the Google Account Creation page (https://accounts.google.com/signup).
2. Fill out the form with your name, username, and password.
3. Provide your birth date, gender, and optionally, your mobile number and alternative email address.
4. Complete the "Prove you're not a robot" verification.
5. Review and accept the Terms of Service and Privacy Policy.
6. Click "Next" or "Create Account" to proceed.
7. Optionally, set up account recovery options.
8. Confirmation that your Google account has been created will be displayed.
9. Optionally, personalize your account settings.
10. Congratulations! You have successfully created a Google account.

These steps assist you in creating a Google account, which can be used for various Google services. If you prefer to use an existing Google account, verify that its associated Google Drive has at least 100 GB of free storage space.

Google Earth Engine (GEE)

GEE is a cloud-based geospatial processing tool with built-in spatial datasets that spans more than 4 decades. A sign-up is required using an active Google account to use the GEE service.

Here is a summary of the steps to set up an Earth Engine project in the Google Cloud Console:

Step 1.

1.   Sign in to the Google Cloud Console: https://console.cloud.google.com.
2.   Create a new project and provide a descriptive name.

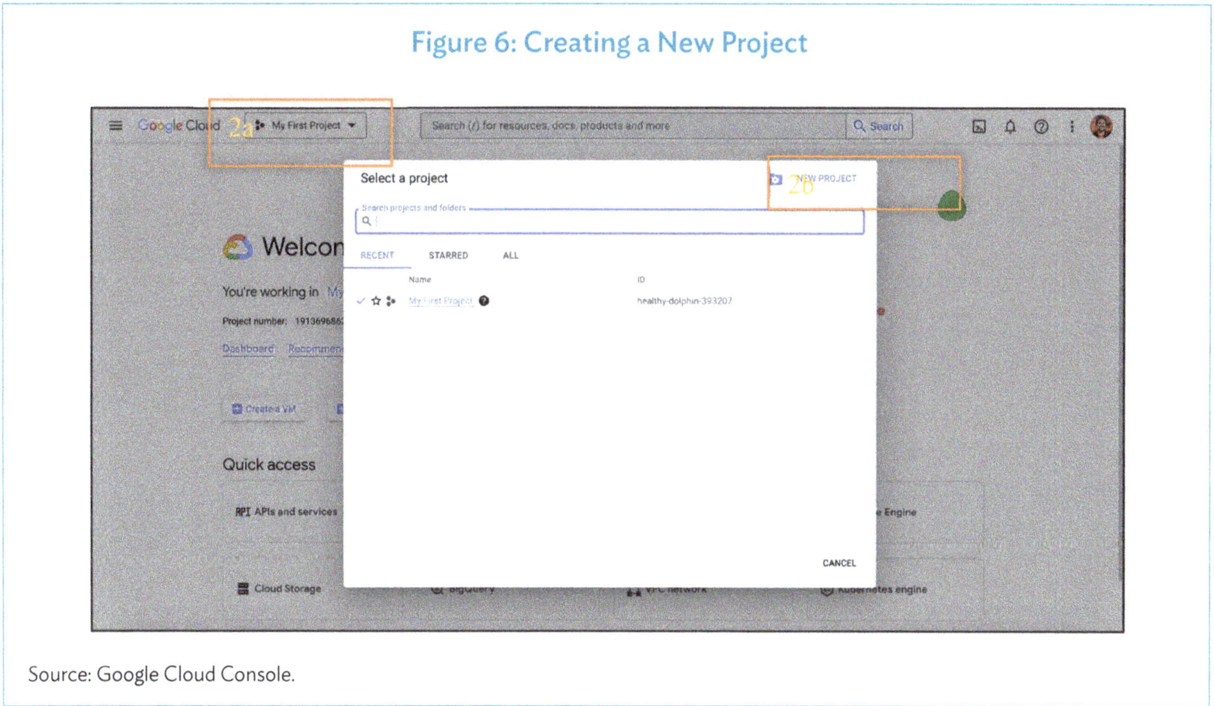

Figure 6: Creating a New Project

Source: Google Cloud Console.

3.   Choose the organization and billing account (if applicable) for the project.
4.   Wait for the project to be created.
5.   Enable the Earth Engine API for your project in the APIs & Services Library.

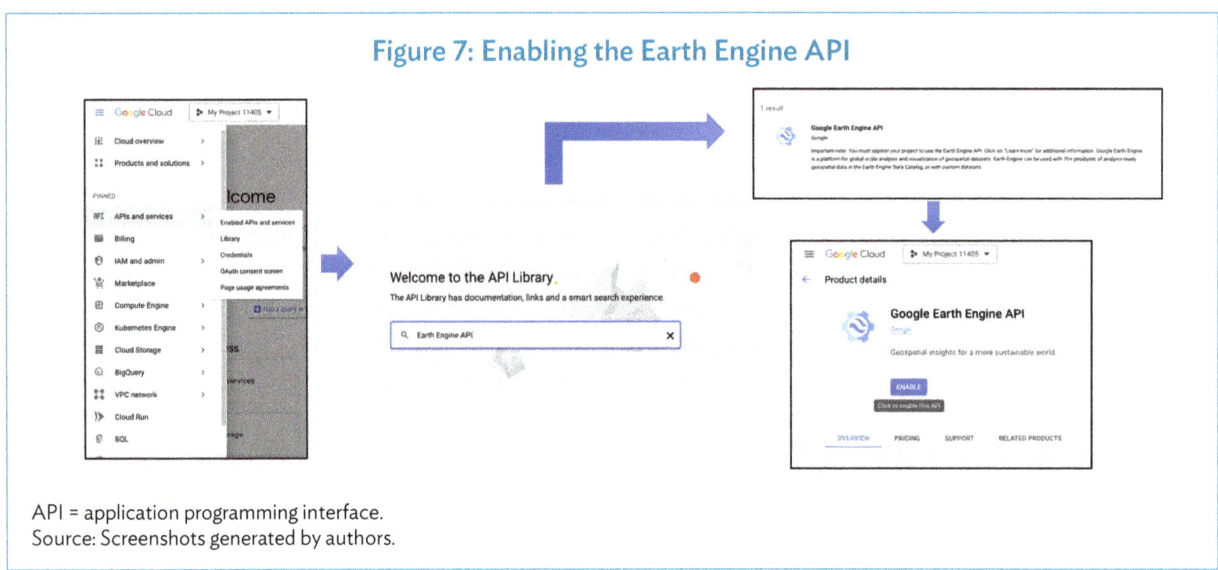

Figure 7: Enabling the Earth Engine API

API = application programming interface.
Source: Screenshots generated by authors.

6.   Set up authentication by creating a service account and downloading the service account key file.
7.   Use the Earth Engine API and the service account key file to configure Earth Engine for your project.

For Step 2, you have two options:

Via Sign Up Form

1.   Go to the Earth Engine website (https://earthengine.google.com/).
2.   Click on "Sign up" or "Get Started" on the homepage (usually located in the upper right of the page).

Figure 8: Getting Started with Earth Engine

Source: Screenshot generated by authors from Google Earth Engine.

Figure 9: Earth Engine Sign-Up Page

Source: Screenshots generated by Authors.

1. Review and accept the Terms of Service and Privacy Policy. Make sure to not skip this step.
2. Complete any additional verification steps if necessary.
3. Submit your application.
4. Wait for Google to review your application.
5. If approved, you'll receive an email with instructions to access and set up your Earth Engine account.
6. Follow the instructions in the email to complete the setup process.
7. Once set up, log in to Earth Engine and start using its features and tools.

Via Cloud Project

In most cases, you do not have an existing Google Cloud Project. If you happen to have an existing Google Cloud Project, you may skip this step.

1. Click "Register a Noncommercial or Commercial Cloud Project."

Figure 10: Registering a Noncommercial or Commercial Project

Source: Screenshots generated by authors.

2.   IMPORTANT: Usage of Earth Engine comes with licensing terms and conditions. Please refer to your organization's policies and Google's Terms and Conditions before selecting the appropriate usage. Failure to comply may result in access to Google Cloud products such as Earth Engine being revoked. For the purpose of this guidebook, we will select Unpaid Usage.

Figure 11: Unpaid Usage for Earth Engine

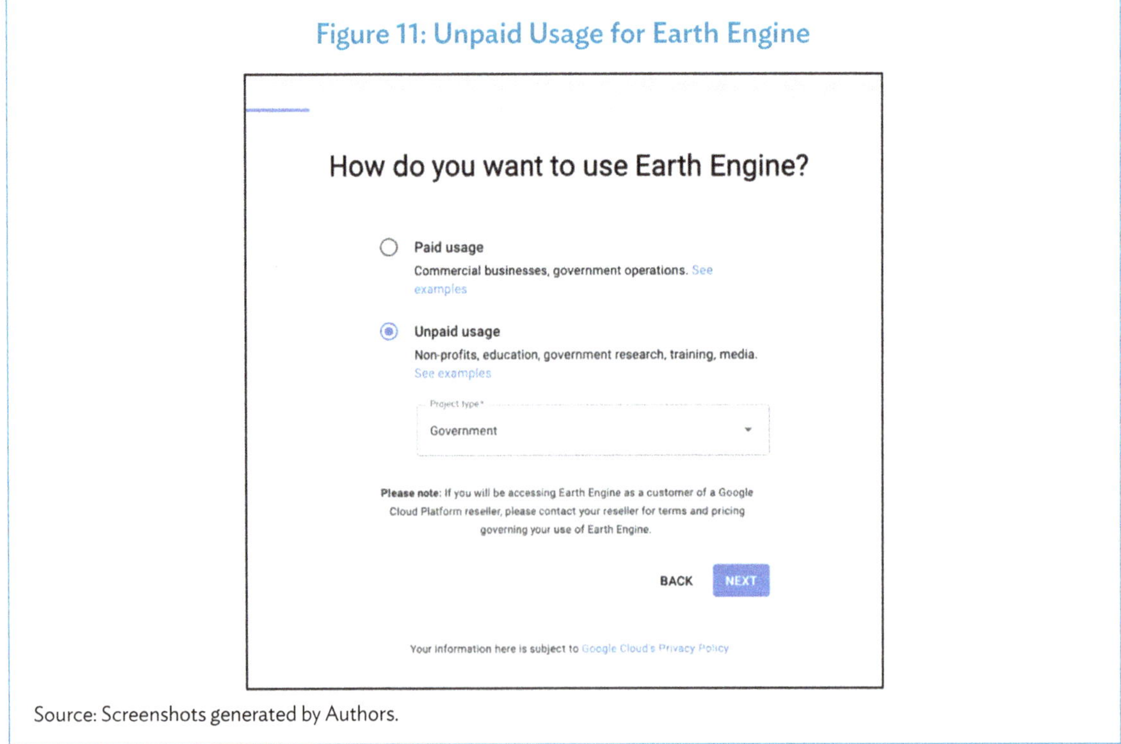

Source: Screenshots generated by Authors.

3. Create a new Google Cloud Project. Again, if you have an existing Google Cloud Project, you may opt to choose it.

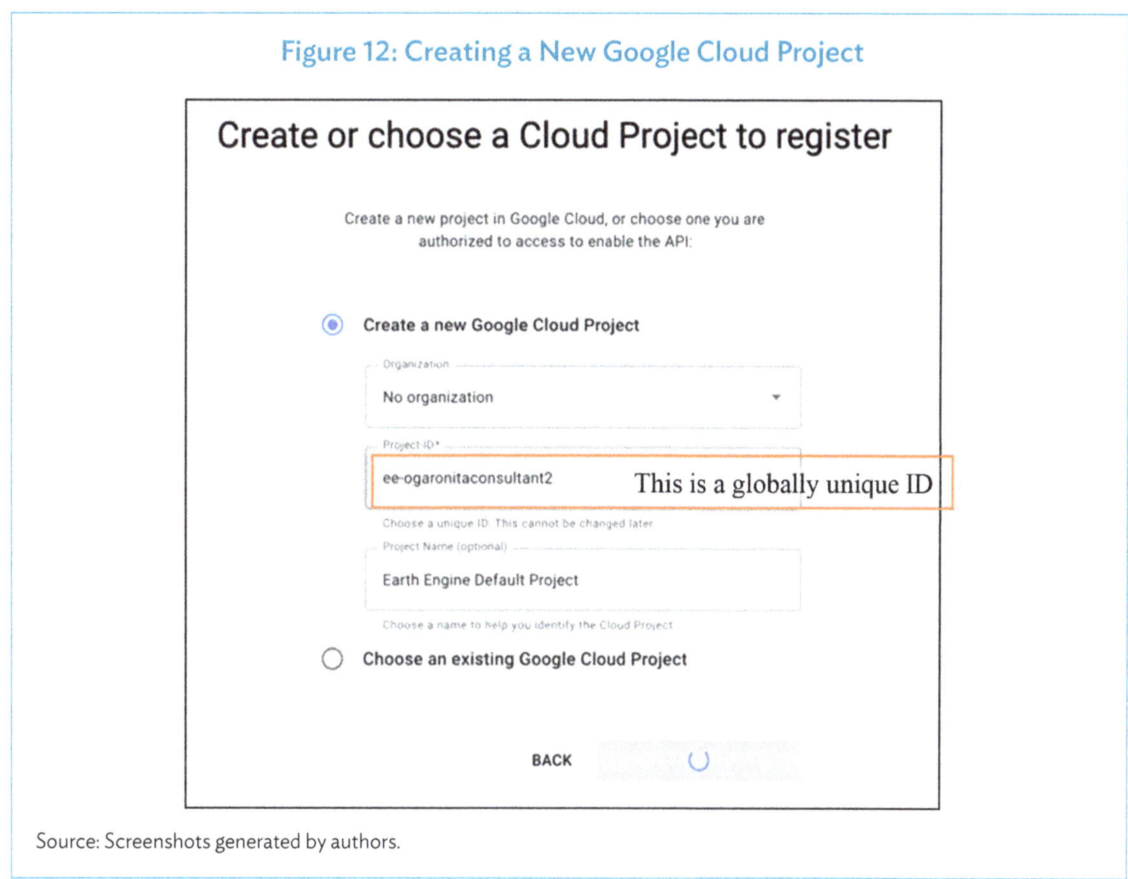

Figure 12: Creating a New Google Cloud Project

Source: Screenshots generated by authors.

4. Once you are done creating and/or selecting a Google Cloud Project, you can now go to Earth Engine.

V. Data Preparation

Acquiring Shapefiles

A shapefile is a simple vector data storage format for storing geographic features' location, shape, and attributes. The geographic features in a shapefile can be represented by points, lines, or polygons. Shapefiles determine the extent of satellite imagery to download. The administrative boundaries of the shapefiles should be consistent with official statistical data.

In this guide, we use QGIS to process shapefile data (alternatively, you can use ArcGIS). Download the US TIGER/Line Road Census shapefile from the US Census Bureau website, which can be found here: https://www2.census.gov/geo/tiger/TIGER2024/PRISECROADS/. The files are in zip file, and make sure to download all of them. Users may opt to use download manager programs/apps for bulk downloading.

Once downloaded, load the files to QGIS by just simply dragging and dropping them into the QGIS Layers panel.

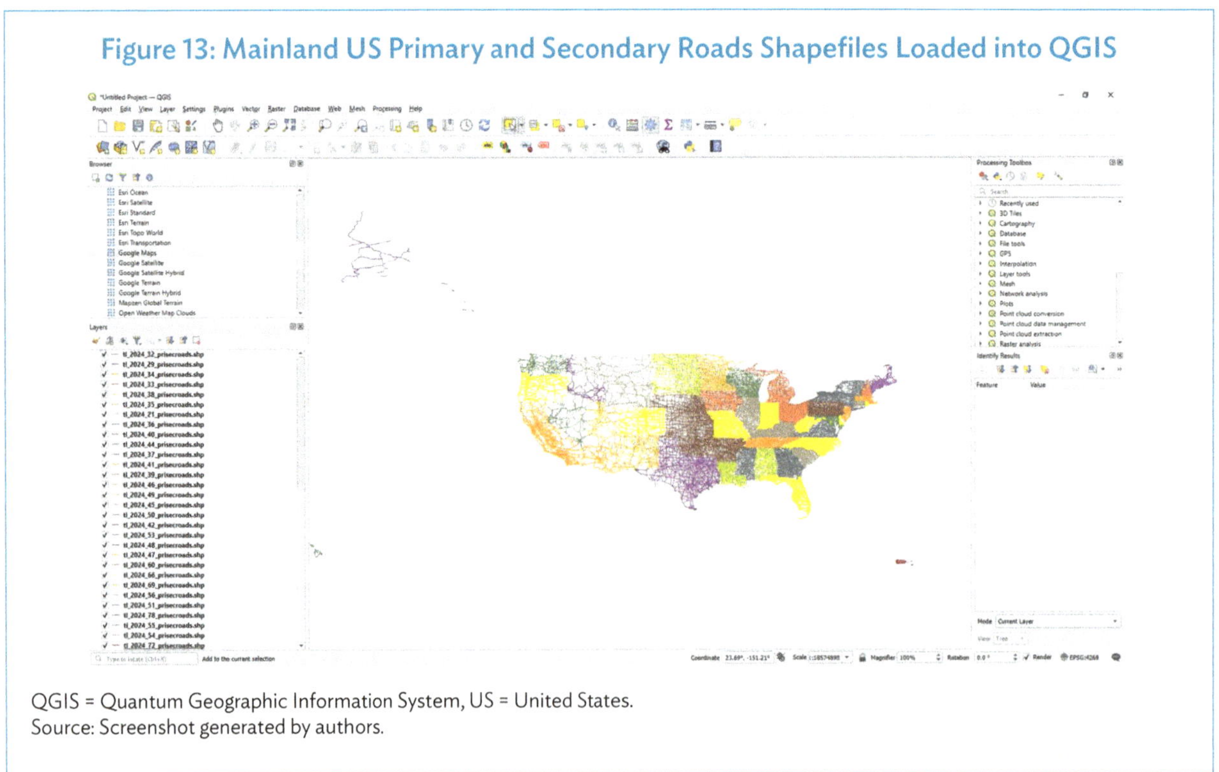

Figure 13: Mainland US Primary and Secondary Roads Shapefiles Loaded into QGIS

QGIS = Quantum Geographic Information System, US = United States.
Source: Screenshot generated by authors.

We need to merge all of the layers into just one layer for us to proceed to our next processes. To do that in QGIS, go to Vector > Data Management Tools > Merge Vector Layers. On the next dialog box, beside Input layers, select the ellipsis button (…), then click Select All. Beside the Merged, click the ellipsis button (…) also, and select which folder you want to save the file. We recommend using the .gpkg format when saving your file. Then, click Run.

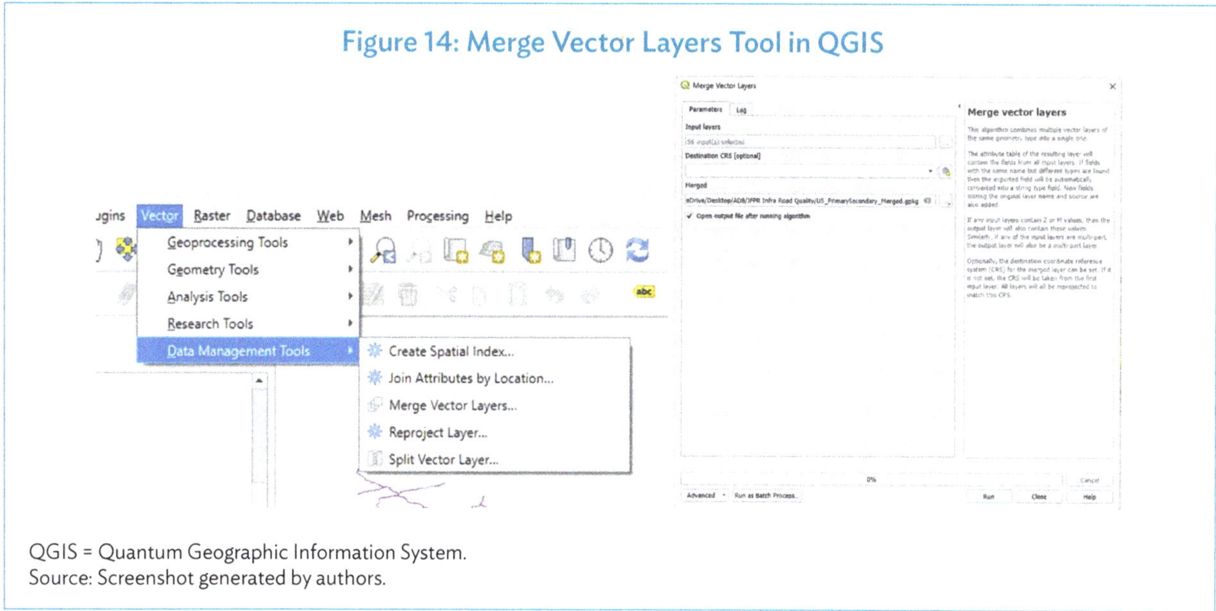

Figure 14: Merge Vector Layers Tool in QGIS

QGIS = Quantum Geographic Information System.
Source: Screenshot generated by authors.

We need to reproject the new file for us to proceed. Under the Processing Toolbox (Processing > Toolbox), search for "Reproject layer." It should be under "Vector general." Under Input layer, select the merged file (in this case, we named it US_PrimarySecondary_Merged). For the Target CRS, click the globe button beside the Target CRS and search for NAD 1983 Equidistant Conic North America.

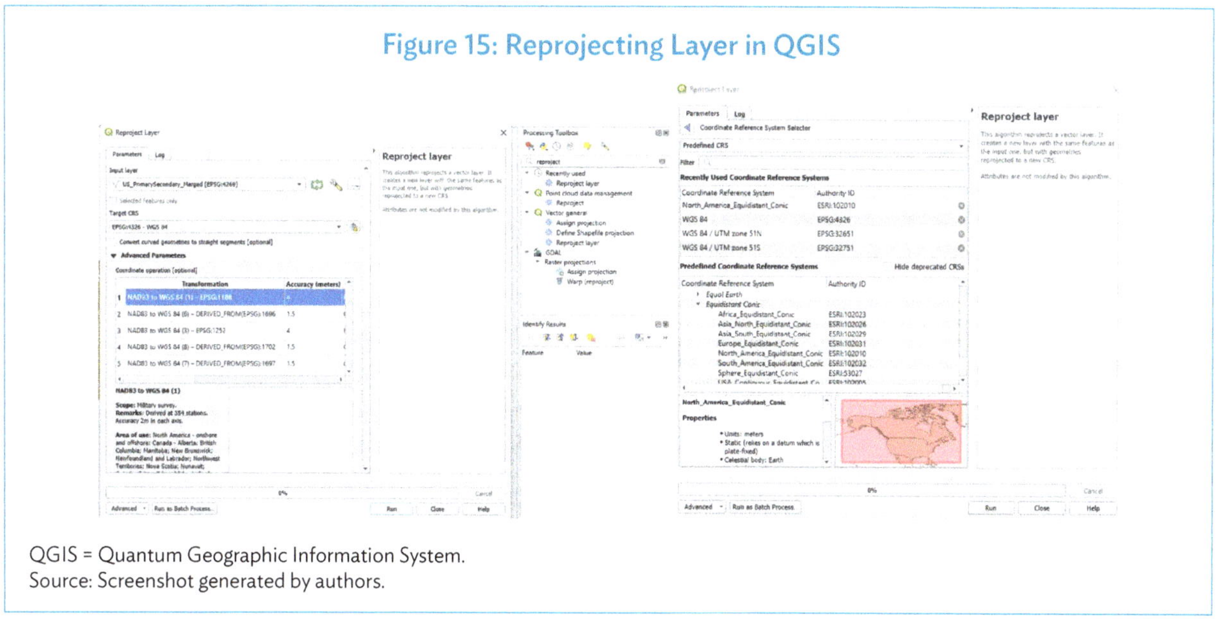

Figure 15: Reprojecting Layer in QGIS

QGIS = Quantum Geographic Information System.
Source: Screenshot generated by authors.

Acquiring Road Bounding Boxes

It is necessary to split roads into 2-kilometer sections in order to obtain our bounding box (or bbox). To do this we use the following procedure:

Step 1.

On the Processing Toolbox panel, search the Points along geometry tool under Vector geometry to **generate points along lines** with Interval spacings of 2,000 meters. For now, let us create a temporary later and we can skip the interpolated points box.

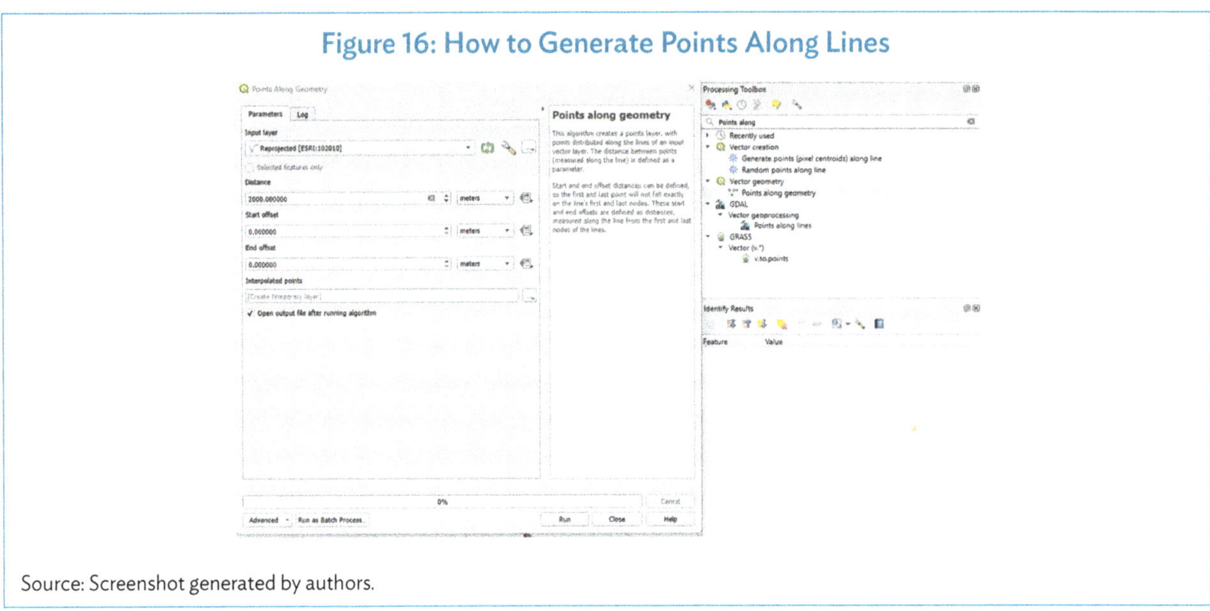

Figure 16: How to Generate Points Along Lines

Source: Screenshot generated by authors.

The resulting feature should appear as shown below:

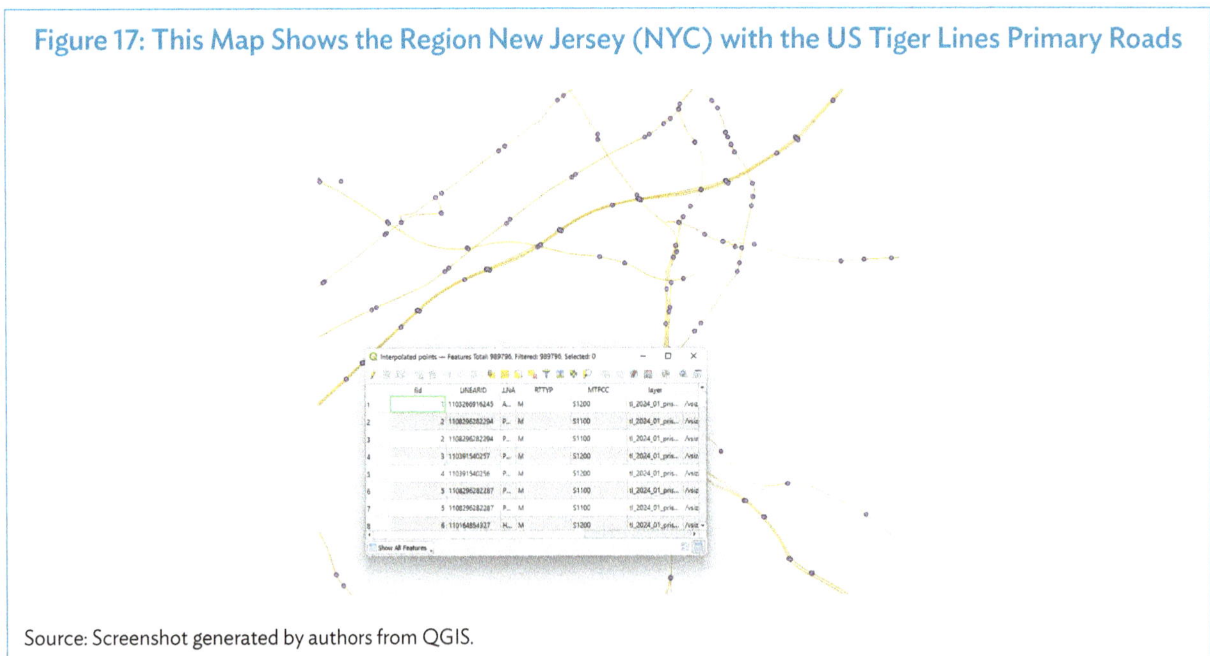

Figure 17: This Map Shows the Region New Jersey (NYC) with the US Tiger Lines Primary Roads

Source: Screenshot generated by authors from QGIS.

Step 2.

We now export the newly created points for processing on Google Colab[16] to create our start points and end points. To do this, we simply export the data as a comma-separated value (CSV):

(**Important:** Make sure to select 'AS_XY' under Layer Options > Geometry. This is so that upon saving as CSV, the latitude and longitude columns will appear on our table.)

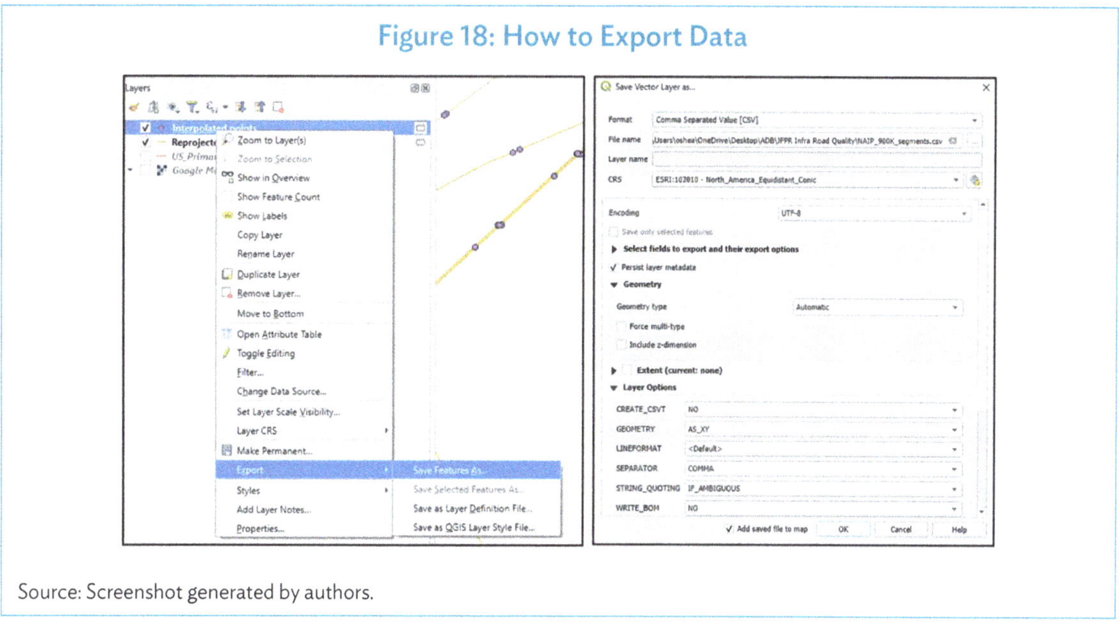

Figure 18: How to Export Data

Source: Screenshot generated by authors.

The exported data now appears in your local directory where you saved it.

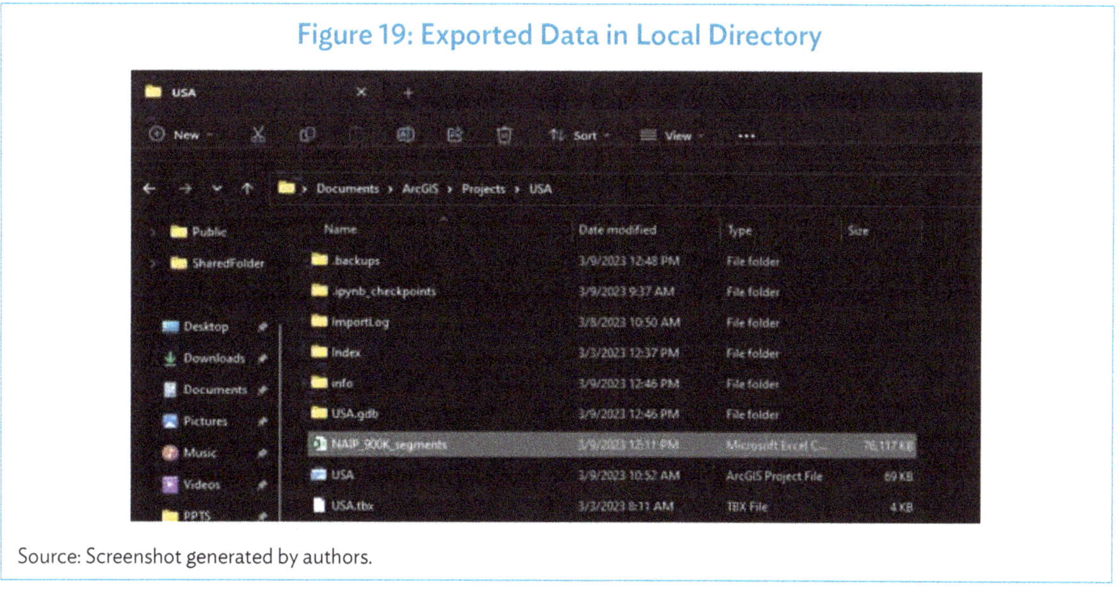

Figure 19: Exported Data in Local Directory

Source: Screenshot generated by authors.

---

[16] Google Colaboratory or Colab is a product from Google Research which allows users to write and execute arbitrary Python code through their own browser.

Uploading the Output to Google Drive

We first must process our bbox before we can begin downloading the imagery, therefore we first upload the newly generated CSV (in our case labeled "**NAIP_900K_segments.csv**") to Google Drive in order to work with it online on Google Colab.

Step 1.

On the Google Chrome address bar, navigate to Google Drive (http://www.drive.google.com/). Follow the prompts to log in to your Google account. Click on the nine dots next to your account and locate the Google Drive icon:

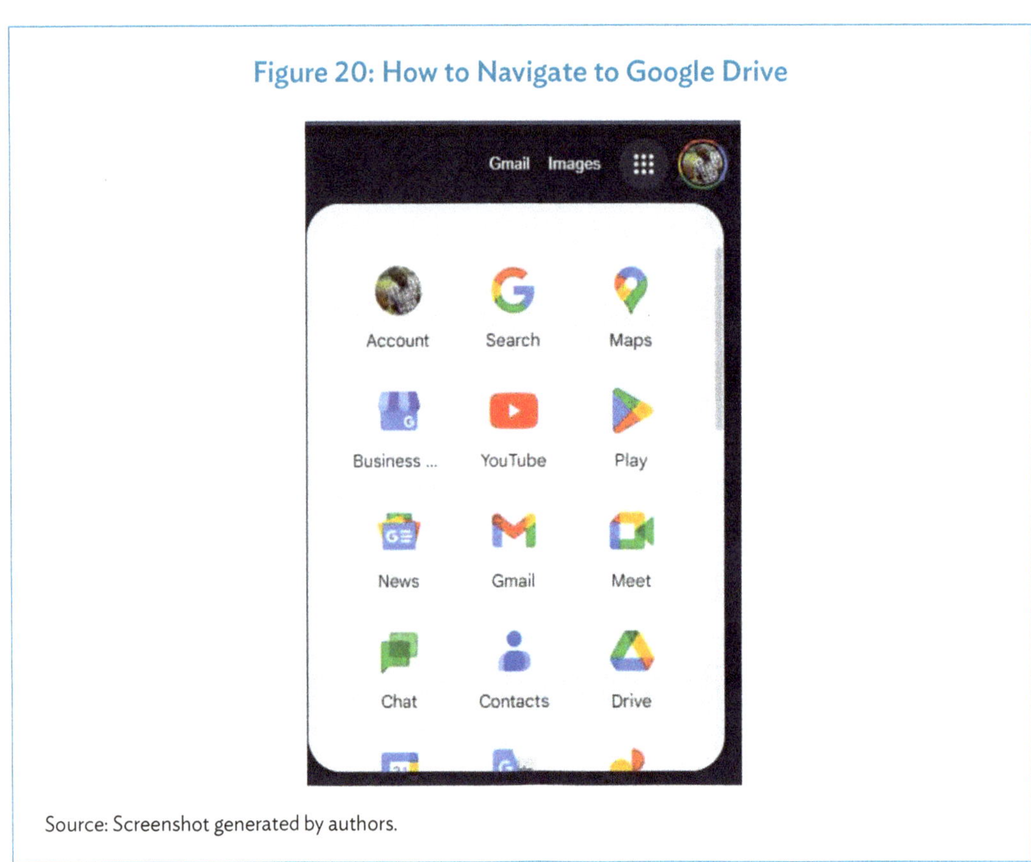

Figure 20: How to Navigate to Google Drive

Source: Screenshot generated by authors.

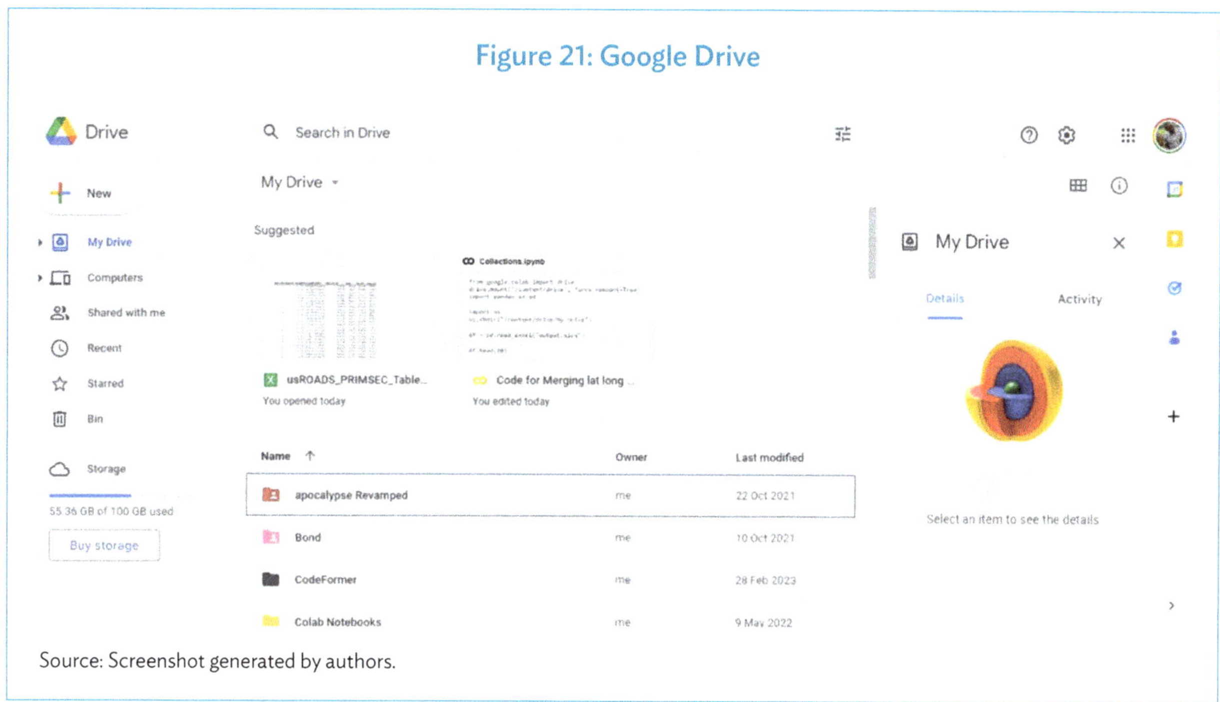

Figure 21: Google Drive

Source: Screenshot generated by authors.

Note: This process will require a considerable amount of storage from your Google Drive, at least 15 GB. Kindly ensure you have sufficient storage. If not, you may consider a Google One (https://one. google.com/about) account. To do this, navigate to Google One plans (https://one.google.com/plans) and upgrade your account to access more storage on your drive. Once this is done, your drive account should have considerably more storage:

Figure 22: Available Cloud Storage

Source: Screenshot generated by authors.

Step 2.

Click on New and then click File Upload.

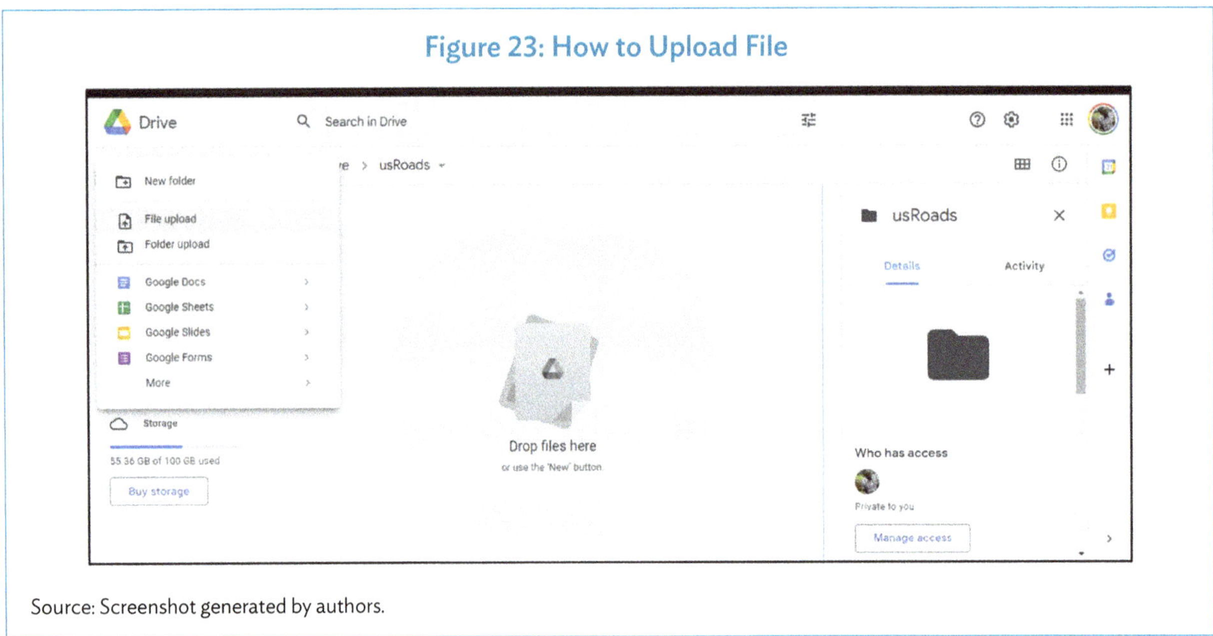

Figure 23: How to Upload File

Source: Screenshot generated by authors.

This is the file needed for generating the bbox and thereafter downloading the satellite imagery of each grid:

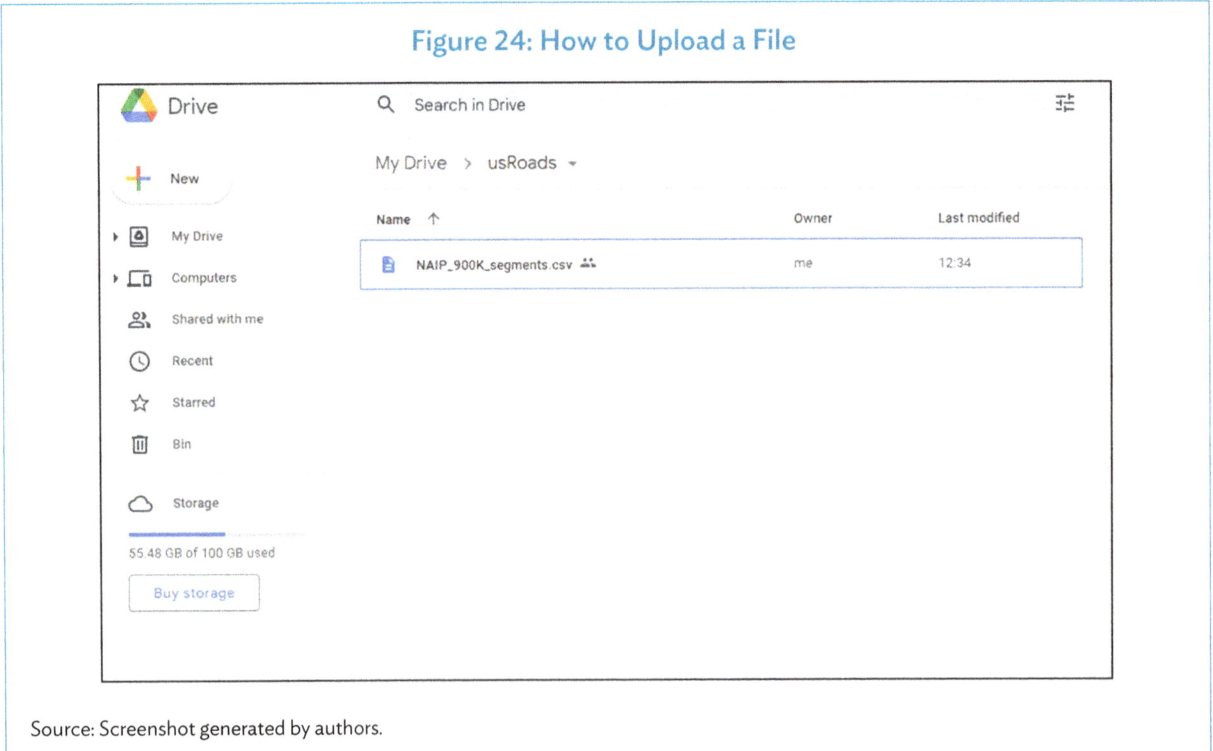

Figure 24: How to Upload a File

Source: Screenshot generated by authors.

Processing Bounding Boxes

Step 1.

Go to Google Colab using the web address https://colab.research.google.com. Make sure to log in to your Google account. Then click Upload.

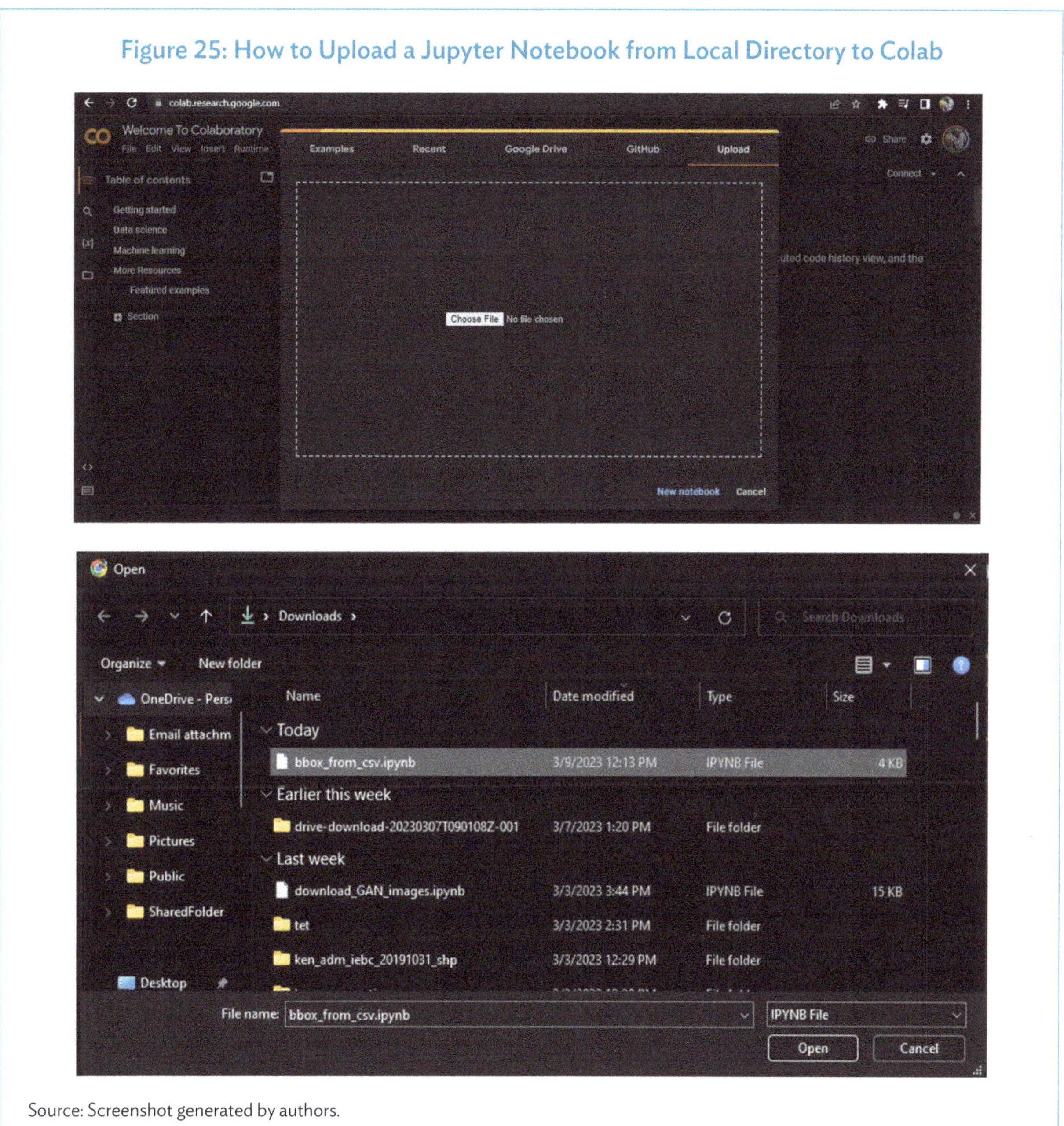

Figure 25: How to Upload a Jupyter Notebook from Local Directory to Colab

Source: Screenshot generated by authors.

Use the notebook "bbox_from_csv.ipynb" and click open.

Step 2.

Click on Connect in the upper right corner.

Figure 26: How to Initialize Colab Environment

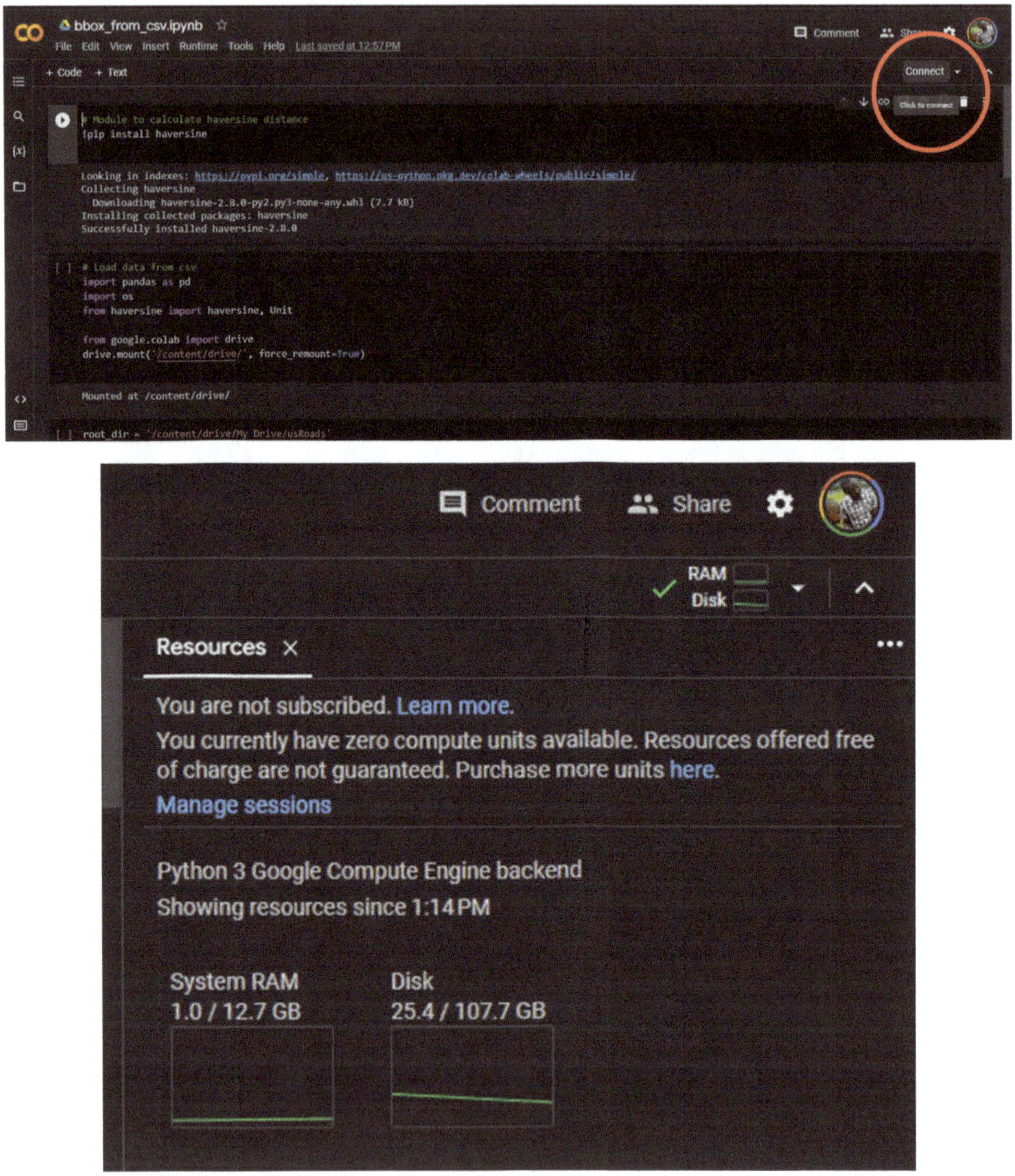

Source: Screenshot generated by authors.

Note: A code cell in a Colab notebook is denoted with square brackets "[]"

Step 3.

To execute, click on each code cell and click the button at the beginning of each code cell.

The first code installs haversine which is used to measure distances:

→   pip install haversine

Code on Installing Haversine

```
[ ]  # Module to calculate haversine distance
     !pip install haversine

     Looking in indexes: https://pypi.org/simple, https://us-python.pkg.dev/colab-wheels/public/simple/
     Collecting haversine
       Downloading haversine-2.8.0-py2.py3-none-any.whl (7.7 kB)
     Installing collected packages: haversine
     Successfully installed haversine-2.8.0
```

The next cell block mounts the Drive:

→   import pandas as pd

→   import os

→   from haversine import haversine, Unit

→   from google.colab import drive

→   drive.mount('/content/drive/', force_remount=True)

Code on Mounting the Drive

```
[ ]  # Load data from csv
     import pandas as pd
     import os
     from haversine import haversine, Unit

     from google.colab import drive
     drive.mount('/content/drive/', force_remount=True)


     Mounted at /content/drive/
```

Step 4.

We will then read the CSV file we just uploaded and convert it to a list:

→   root_dir = 'path to your folder of US Roads'

→   infile = 'NAIP_900K_segments.csv'

→   outfile = 'bbox_ccords_NAIP_900k.csv'

→   df = pd.read_csv(os.path.join(root_dir,infile), encoding="latin-1")

→   data = df.values.tolist()

Code on Reading the CSV Files

```python
root_dir = '/content/drive/My Drive/usRoads'
infile = 'NAIP_900K_segments.csv'
outfile = 'bbox_coords_NAIP_900k.csv'

df = pd.read_csv(os.path.join(root_dir,infile), encoding="latin-1")
data = df.values.tolist()
```

Step 5.

The comments on the code cell block highlight each operation that is occurring by running the following code:

→   output = [['X', 'Y', 'fid', 'LINEARID', 'FULLNAME', 'RTTYP', 'MTFCC', 'distance', 'angle', 'LAT_ END', 'LON_END']]

→   seg_id = 1

→   for ix in range(0,len(data)-1):
        if data[ix + 1][3] == data[ix][3]:
            dist = haversine((data[ix][1], data[ix][0]), (data[ix + 1][1], data[ix + 1][0]))
            row = data[ix][0:2] + [data[ix][2], seg_id, data[ix][4], data[ix][5], data[ix][6], dist, 0,
            data[ix + 1][1], data[ix + 1][0]]
            output.append(row)
            seg_id += 1
        else:
            seg_id = 1

→   df_out = pd.DataFrame(output[1:], columns = output[0])

Code on Creating the Bounding Box

```python
# The .csv file contains roads identified by LINEARID.
# Each road is in turn divided into 2-km segments
# Create a bounding box of lat and long coords for each segment

# Initialize output list
output = [['X', 'Y', 'fid', 'LINEARID', 'FULLNAME', 'RTTYP', 'MTFCC', 'distance', 'angle', 'LAT_END', 'LON_END']]
seg_id = 1

# Loop until second-to-last observation
for ix in range(0, len(data) - 1):
    # Add end-point lat and long if the following point is part of the same road
    if data[ix + 1][3] == data[ix][3]:  # Checks if LINEARID is the same
        dist = haversine((data[ix][1], data[ix][0]), (data[ix + 1][1], data[ix + 1][0]))
        # Append the row with exactly 11 columns
        row = data[ix][0:2] + [data[ix][2], seg_id, data[ix][4], data[ix][5], data[ix][6], dist, 0, data[ix + 1][1], data[ix + 1][0]]
        output.append(row)
        seg_id += 1
    else:
        # This is the last segment of a road. Skip and reset counter
        seg_id = 1

# Now write the output list to a csv file and save it within your root dir
df_out = pd.DataFrame(output[1:], columns = output[0])
```

Step 6.

The shape returns the length of our CSV file, the output file has been given the name "**bbox_coords_NAIP_900k.csv**" and should be visible within the same folder where our previous CSV file was:

→   df_out.to_csv(os.path.join(root_dir, outfile))

→   df_out.shape[0]

Code on Saving the Output to CSV

```python
df_out.to_csv(os.path.join(root_dir, outfile))
df_out.shape[0]

842958
```

Downloading Satellite Imagery

Step 1.

Using the procedure in the *Processing Bounding Box* section, we now upload the code "**download_GAN_images.ipynb**" which downloads our satellite imagery.

Just as in the above section, click on connect to initialize the Colab resources.

Step 2.

→   The first block downloads the necessary packages needed to download the satellite images:

→   !pip install geemap

→   !pip install retry

→ !pip install haversine

Code on Installing Geemap, Retry, and Haversine

```
#install neccesary packages
!pip install geemap
!pip install retry
!pip install haversine
```

Step 3.

This block initializes some pre-installed modules plus the ones we just downloaded, initializes the Earth Engine API with the high-volume end point, and mounts to your Google Drive:

→ Import ee
→ Import os
→ Import pandas as pd
→ Import time
→ Import logging
→ Import geemap
→ From multiprocessing import Pool
→ From haversine import haversine, Unit
→ ee.Authenticate()
→ ee.Initialize(opt_url= 'https://earthengine-highvolume.googleapis.com')
→ from google.colab import drive
→ drive.mount('/content/drive', force_remount=True)

Code on Initializing Modules and Earth Engine API

```
# utils
import ee
import os
import pandas as pd
import time
#from retry import retry
import logging
import geemap
from multiprocessing import Pool
from haversine import haversine, Unit

ee.Authenticate()
ee.Initialize(opt_url='https://earthengine-highvolume.googleapis.com')

from google.colab import drive
drive.mount('/content/drive', force_remount=True)
```

Step 4.

The comments on the code cell block highlight each operation that is occurring by running the following code:

- → root_dir = 'path to your folder of US Roads'
- → sentinel_dir = 'path to your folder of low-resolution US Roads'
- → naip_dir = 'path to your folder of high-resolution US Roads'
- → date_start = 'target start date of imagery'
- → date_end = 'target end date of imagery'
- → lr_scale = 10
- → scale_factor = 4

Code on Necessary Specifics

```python
# change to root dir
root_dir = '/content/drive/My Drive/usRoads'
sentinel_dir =  '/content/drive/My Drive/usRoads/LQ'
naip_dir = '/content/drive/My Drive/usRoads/GT'


# Key variables
date_start = '2020-06-01'
date_end = '2020-12-30'


# Set lower resolution image scale
# For sentinel 10 meters per pixel
lr_scale = 10


# The ratio in scale between low-res and high-res images
# NAIP data are 1 meter per pixel and SENTINEL 10 meters per pixel
# The maximum scale factor is 10
scale_factor = 4
```

Specify directory variables and some pixel-scaling variables:

- → csv_file = 'bbox_coords_NAIP_900k.csv'
- → df = pd.read_csv(os.path.join(root_dir,csv_file))
- → df.drop(["Unnamed: 0", "ORIG_FID", "LINEARID", "SEG_ID", "MTFC", "FULLNAME"], inplace = True, axis = 1)
- → data_inputs = df.values.tolist()
- → print(data_inputs[0])

Code on Specifying Directory Variables

```python
# Read the csv file, get bounding box coordinates, and convert to list
# The coordinates are [lat,lon,lat,lon]
csv_file = 'bbox_coords_NAIP_900k.csv'

df = pd.read_csv(os.path.join(root_dir,csv_file))

df.drop(["Unnamed: 0","ORIG_FID","LINEARID","SEG_ID", "MTFC", "FULLNAME"], inplace = True, axis = 1)
# df.drop(["LINEARID", "SECTIONID", "FULLNAME"], inplace = True, axis = 1)
data_inputs = df.values.tolist()
print(data_inputs[0])
```

```
[1.0, 29.56932259, -98.33449554, 29.58189201, -98.34494019, 1.7244385968366662]
```

Read the imported CSV file and drop non-essential columns:

→   for x in data_inputs[0:5]:
        distance = x[5]
→   print("distance in km: ", distance)

Code on Calculating Haversine Distance

```python
# Calculate haversine distance
# Specify inputs as (lat,lon)
for x in data_inputs[0:5]:
    distance = x[5]
    # a=(x[1],x[2])
    # b=(x[3],x[4])
    # print("distance in km: ", haversine(a,b))
    print("distance in km: ", distance)
```

Step 5.

The comments on the code cell block highlight each operation that is occurring:

→   os.chdir(sentinel_dir)
→   len(os.listdir())
→   def maskS2clouds(image):
        qa = image.select('QA60');
        cloudBitMask = 1 << 10;
        cirrusBitMask = 1 << 11;
        mask = qa.bitwiseAnd(cloudBitMask).eq(0)
        mask = mask.And(mask.bitwiseAnd(cirrusBitMask).eq(0))
        return image.updateMask(mask).divide(10000)

Code on Functions to Mask Clouds

```python
# Shift to sentinel directory
os.chdir(sentinel_dir)

# ensure the directory is empty
len(os.listdir())

0

# Function to mask clouds
def maskS2clouds(image):
    qa = image.select('QA60');

    #Bits 10 and 11 are clouds and cirrus, respectively.
    cloudBitMask = 1 << 10;
    cirrusBitMask = 1 << 11;

    #Both flags should be set to zero, indicating clear conditions.
    mask = qa.bitwiseAnd(cloudBitMask).eq(0)
    mask =  mask.And(mask.bitwiseAnd(cirrusBitMask).eq(0))

    return image.updateMask(mask).divide(10000)
```

Step 6.

To download sentinel images (low resolution), run the function below:

```python
def download_road_images2(road_sec):
    bbox = [road_sec[2], road_sec[1], road_sec[4], road_sec[3]]
    aoi = ee.Geometry.Rectangle(bbox)
    sentinel = (ee.ImageCollection('COPERNICUS/S2_SR')
      .filterBounds(aoi)
      .filterDate(date_start,date_end)
      .filter(ee.Filter.lt('CLOUDY_PIXEL_PERCENTAGE', 2)).map(maskS2clouds)
    )
    a = (road_sec[1], road_sec[2])
    b = (road_sec[1], road_sec[4])
    width = haversine(a, b)
    px_width = int(width*1000/lr_scale)
if __name__ == '__main__':
    pool = Pool(25)
    pool.map(download_road_images2, data_inputs)
    poll.close()
    pool.terminate()
```

Code on Downloading Sentinel Images

```python
# function download Sentinel images
def download_road_images2(road_sec):
    bbox = [road_sec[2], road_sec[1], road_sec[4], road_sec[3]] #your bounding box containing coordinates
    aoi = ee.Geometry.Rectangle(bbox) #Geometry being a rectangle
    # Specify your sentinel image collection
    sentinel = (ee.ImageCollection('COPERNICUS/S2_SR')
            .filterBounds(aoi)
            .filterDate(date_start,date_end)
            .filter(ee.Filter.lt('CLOUDY_PIXEL_PERCENTAGE', 2)).map(maskS2clouds)

    )

    # Get max of width and height of bounding box, and convert to
    # pixels based on scale
    # Distance is returned in km

    # Width (diff between longitudes)
    a = (road_sec[1], road_sec[2])
    b = (road_sec[1], road_sec[4])
    width = haversine(a, b)
    px_width = int(width*1000/lr_scale)
```

```python
if __name__ == '__main__':
    pool = Pool(25)
    pool.map(download_road_images2, data_inputs)

    pool.close()
    pool.terminate()
```

Step 7.

Shift to the NAIP directory:

→   os.chdir(naip_dir)

→   len(os.listdir())

Code on Shifting to NAIP Directory

```python
[ ] os.chdir(naip_dir)

    len(os.listdir())

    0
```

Step 8.

Download NAIP imagery with the function below:

```
a = (road_sec[1], road_sec [2])
b = (road_sec[3], road_sec[2])
height = haversine(a,b)
px_height = int(height*1000/lr_scale)
dim = str(px_width)+'x'+str(px_height)
if (px_width > 64 and px_height > 64):
    if sentinel.size().getInfo() > 0:
        image = sentinel.median().clip(aoi)
            vis_params = {'min': 0.0, 'max': 0.3, 'bands': ['B4', 'B3', 'B2'],}
        x = f"{str(road_sec[0])}.jpg"
        tsk = geemap.get_image_thumbnail(image, x, vis_params, dimensions=dim, region= aoi,
        format='jpg')
def download_road_images(road_sec):
    bbox = [road_sec[2], road_sec[1], road_sec[4], road_sec[3]]
    aoi = ee.Geometry.Rectangle(bbox)
    naip = (ee.ImageCollection('USDA/NAIP/DOQQ')
        .filterBounds(aoi)
        .filterDate(date_start,date_end)
    )
    a = (road_sec[1], road_sec[2])
    b = (road_sec[1], road_sec[4])
    width = haversine(a, b)
    px_width = (int(width*1000/lr_scale))*scale_factor
if __name__ == '__main__':
    logging.basicConfig()
    pool = Pool(25)
    pool.map(download_road_images, data_inputs[0:2000])
    pool.close()
    pool.terminate()
a = (road_sec[1], road_sec[2])
b = (road_sec[3], road_sec[2])
height = haversine(a,b)
px_height = (int(height*1000/lr_scale))*scale_factor
dim = str(px_width)+'x'+str(px_height)
```

```
→    if (px_width/scale_factor > 64 and px_height/scale_factor > 64):
        if naip.size().getInfo() > 0:
            image = naip.median()
            vis_params = { 'bands': ['R', 'G', 'B'], 'min': 0, 'max': 255}
        x = f"{str(road_sec[0])}.jpg"
        tsk = geemap.get_image_thumbnail(image, x, vis_parmas, dimensions = dim, region = aoi,
        format = 'jpg')
```

Code on Downloading NAIP Imagery

```python
# Height (diff between latitudes)
a = (road_sec[1], road_sec[2])
b = (road_sec[3], road_sec[2])
height = haversine(a,b)
px_height = int(height*1000/lr_scale)

dim = str(px_width)+'x'+str(px_height)

# Do not download images that are less than 64 pixels in any dimension
# ESRGAN will not accept small images
if (px_width > 64 and px_height > 64):

    if sentinel.size().getInfo() > 0:
        image = sentinel.median().clip(aoi)
        vis_params  = {
                    'min': 0.0,
                    'max': 0.3,
                    'bands': ['B4', 'B3', 'B2'],
                }

        x =  f"{str(road_sec[0])}.jpg"
        # Fetch the URL from which to download the image.
        tsk = geemap.get_image_thumbnail(image, x, vis_params, dimensions=dim, region= aoi, format='jpg')
```

```python
# You can add a flag depending on the number of files yoy want to install on the data_inputs part
#  eg: data_inputs[1000] if you are downloading 1000 files.

def download_road_images(road_sec):
    bbox = [road_sec[2], road_sec[1], road_sec[4], road_sec[3]]
    aoi = ee.Geometry.Rectangle(bbox)
    naip = (ee.ImageCollection('USDA/NAIP/DOQQ')
            .filterBounds(aoi)
            .filterDate(date_start,date_end)
    )

    # Get max of width and height of bounding box, and convert to
    # pixels based on scale
    # For NAIP, scale is 1m per pixel
    # Distance is returned in km

    # Width (diff between longitudes)
    a = (road_sec[1], road_sec[2])
    b = (road_sec[1], road_sec[4])
    width = haversine(a, b)
    px_width = (int(width*1000/lr_scale))*scale_factor
```

```python
if __name__ == '__main__':
    logging.basicConfig()
    pool = Pool(25)
    pool.map(download_road_images, data_inputs[0:2000])

    pool.close()
    pool.terminate()
```

```python
[ ]    # Height (diff between latitudes)
       a = (road_sec[1], road_sec[2])
       b = (road_sec[3], road_sec[2])
       height = haversine(a,b)
       px_height = (int(height*1000/lr_scale))*scale_factor

       dim = str(px_width)+'x'+str(px_height)

       # Only download images also downloaded for low-res data
       # Do not download images that are less than 64 pixels in any dimension
       # ESRGAN will not accept small images
       if (px_width/scale_factor > 64 and px_height/scale_factor > 64):
         if naip.size().getInfo() > 0:
           image = naip.median()
           vis_params = {
             'bands': ['R', 'G', 'B'],
             'min': 0,
             'max': 255
           }

           x = f"{str(road_sec[0])}.jpg"
           # Fetch the URL from which to download the image.
           tsk = geemap.get_image_thumbnail(image, x, vis_params, dimensions = dim, region= aoi, format='jpg')
```

Step 9.

Assert file since the training code requires your pair data to have the same length of file sizes:

→   import os

→   import shutil

→   import re

→   common_files = list(set.intersection(set(os.listdir(naip_dir)), set(os.listdir(sentinel_dir))))

→   print(len(common_files))

Code on Asserting Files

```
▼ FILE UTILITIES

                                                        + Code    + Text

[ ]  # Import relevant libraries
     import os
     import shutil
     import re


[ ]  # Get common files in NAIP and SENTINEL directories
     common_files = list(set.intersection(set(os.listdir(naip_dir)), set(os.listdir(sentinel_dir))))
     print(len(common_files))
```

Step 10.

Now move the pair Sentinel and NAIP common files to the dataset folder in your SRGAN training code.

→   dir_gt = root_dir+ '/Real-ESRGAN/datasets/METRICS/GT'
→   dir_lq = root_dir+ '/Real-ESRGAN/datasets/METRICS/LQ'
→   if not os.path.exists(dir_gt):
        os.makedirs(dir_gt)
→   if not os.path.exists(dir_lq):
        os.makedirs(dir_lq)
→   for fname in common_files:
        shutil.move(naip_dir+'/'+fname, dir_gt)
→   for fname in common_files:
        shutil.move(sentinel_dir+'/'+fname, dir_lq)

Code on Moving Sentinel and NAIP Common Files to the Dataset Folder

```python
# Create relevant directories and copy files
# For pairwise ESRGAN training, ensure that both high quality
# and low quality images are stored at the correct path,
# which must match the information in the configuration file
# finetune_realesrgan_x4plus_pairdata.yml.

# Create relevant directories
dir_gt = root_dir+'/Real-ESRGAN/datasets/METRICS/GT'
dir_lq = root_dir+'/Real-ESRGAN/datasets/METRICS/LQ'


if not os.path.exists(dir_gt):
    os.makedirs(dir_gt)

if not os.path.exists(dir_lq):
    os.makedirs(dir_lq)

# Move common high-res files
for fname in common_files:
  shutil.move(naip_dir+'/'+fname, dir_gt)

# Move common low-res files

for fname in common_files:
  shutil.move(sentinel_dir+'/'+fname, dir_lq)
```

Step 11.

Now assert the file sizes:

→   len(os.listdir(dir_gt)), len(os.listdir(dir_lq))

Code on Asserting File Sizes

```
[ ]  len(os.listdir(dir_gt)) , len(os.listdir(dir_lq))
```

VI. Training the Super-Resolution Model (REAL-ESRGAN)

Using the procedure in the *Processing Bounding Boxes* section, use the code "Train-OP.ipynb" which is used to fine-tune the pre-trained GAN model.

Real-ESRGAN and Basic-SR

Basic Super Restoration (Basic SR) is an open-source image and video restoration toolbox based on PyTorch that includes tools for super-resolution, denoising, deblurring, and JPEG artifact removal. It also includes high-level classes that handle training and generation of basic Generative Adversarial network architectures such as the Generator and the Discriminator.

The Real-ESRGAN architecture utilizes the Basic SR pipeline that encompasses all tools from model generation to training.

Clone the GitHub repository from this site (https://github.com/xinntao/Real-ESRGAN).

- If everything is done via Colab, clone it onto a local Google Drive directory and not on the Colab virtual machine.

  Note: This process should only be done once.

We run the following list of commands to install the necessary prerequisites required by the Real-ESRGAN module:

→ !pip install basicr

→ !pip install -r requirements.txt

→ !python setup.py develop

Code on Installing Necessary Prerequisites for the ESRGAN Module

```
RUN

#These are the necessary requirements that allow the basic-sr api to function properly
!pip install basicsr
!pip install -r requirements.txt
#This will create the necessary environment with which we can train our AI model
!python setup.py develop
```

To create the "metainfo.txt" configuration file, we then run the next sequence of commands in order to generate a configuration file that will be used in the training of the super resolution model:

→   !python scripts/generate_meta_info_pairdata.py –input ../DEMO/demo/GT ../DEMO/demo/LQ --meta_info ../DEMO/demo/meta_info/meta_info_GUN_sub_pair.txt

Code on Creating "metainfo.txt" Configure File

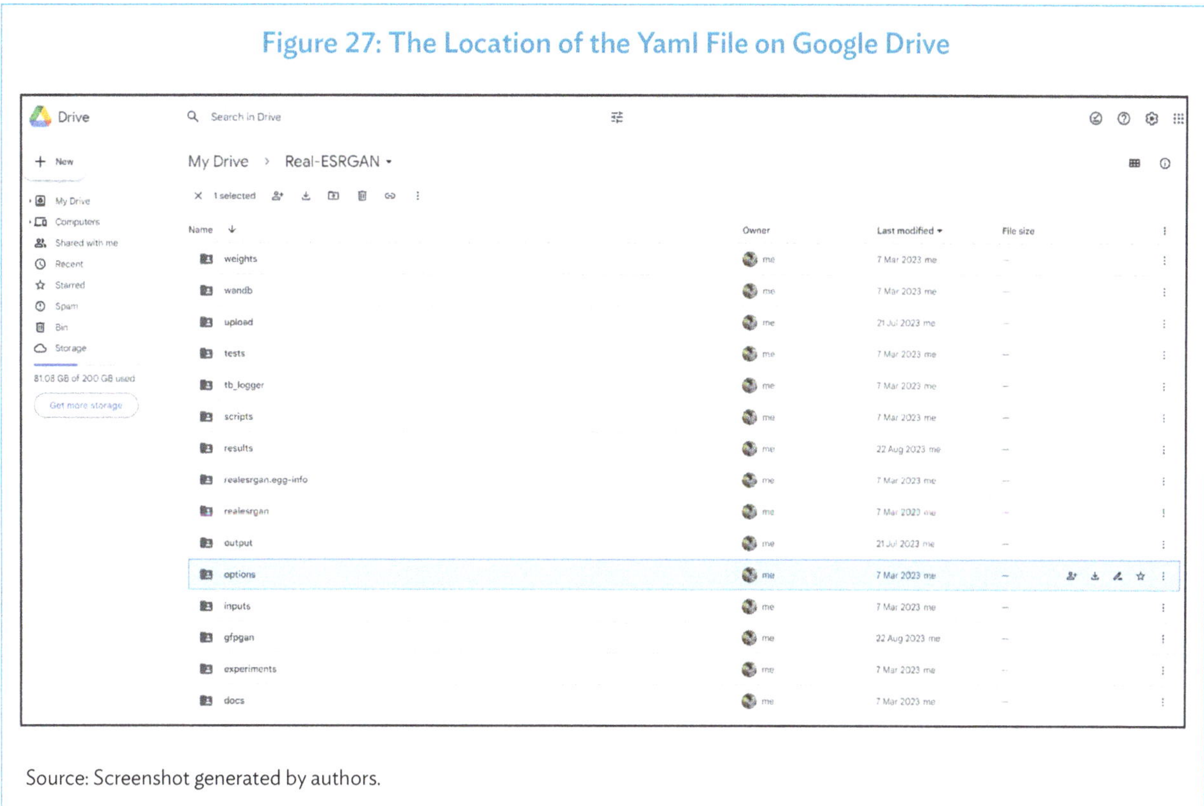

In the Real-ESRGAN folder within our drive, we adjust another configuration Yaml file for training -> finetune_realesrgan_x4plus_pairdata.yml which is in the options folder:

Figure 27: The Location of the Yaml File on Google Drive

Source: Screenshot generated by authors.

Figure 28: Editing the Yaml File

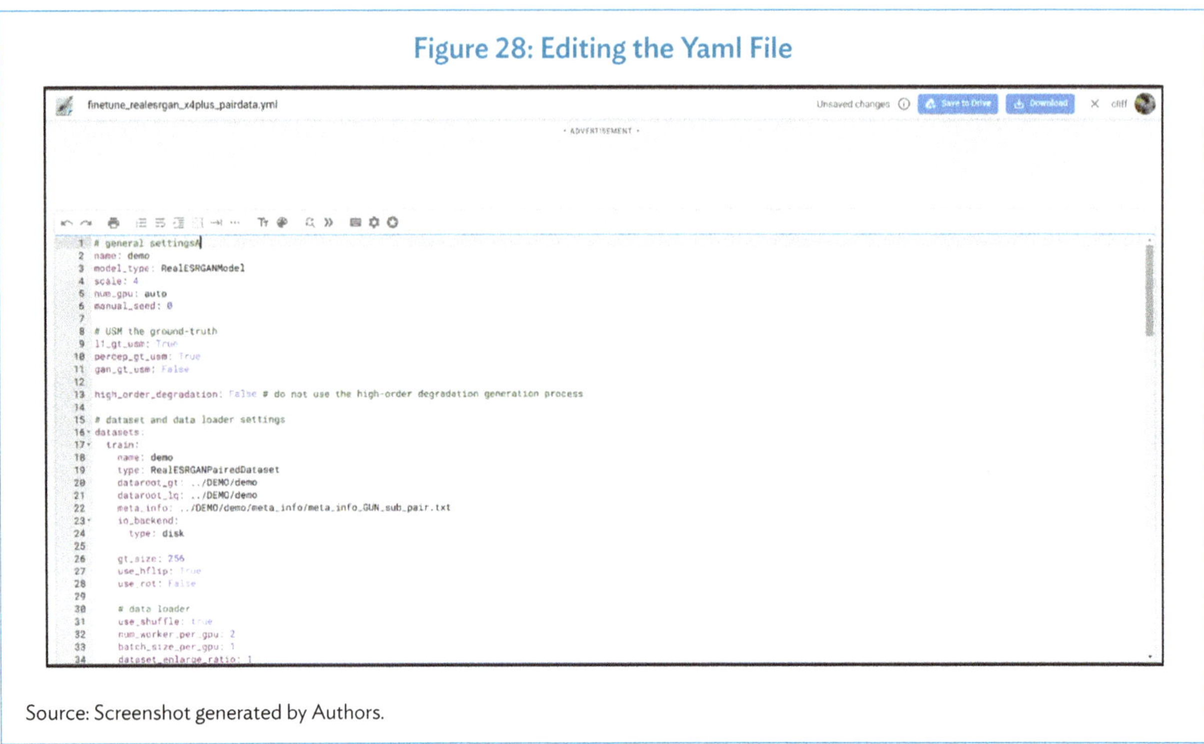

Source: Screenshot generated by Authors.

Download the relevant Real-ESRGAN releases, RealESRGAN_X4_plus.pth and ReaESRGAN_X4_plus_NetD.pth-> https://github.com/xinntao/Real-ESRGAN/releases, and place in the experiments pretrained folder:

Figure 29: Location Where the Pretrained Models Should be Placed

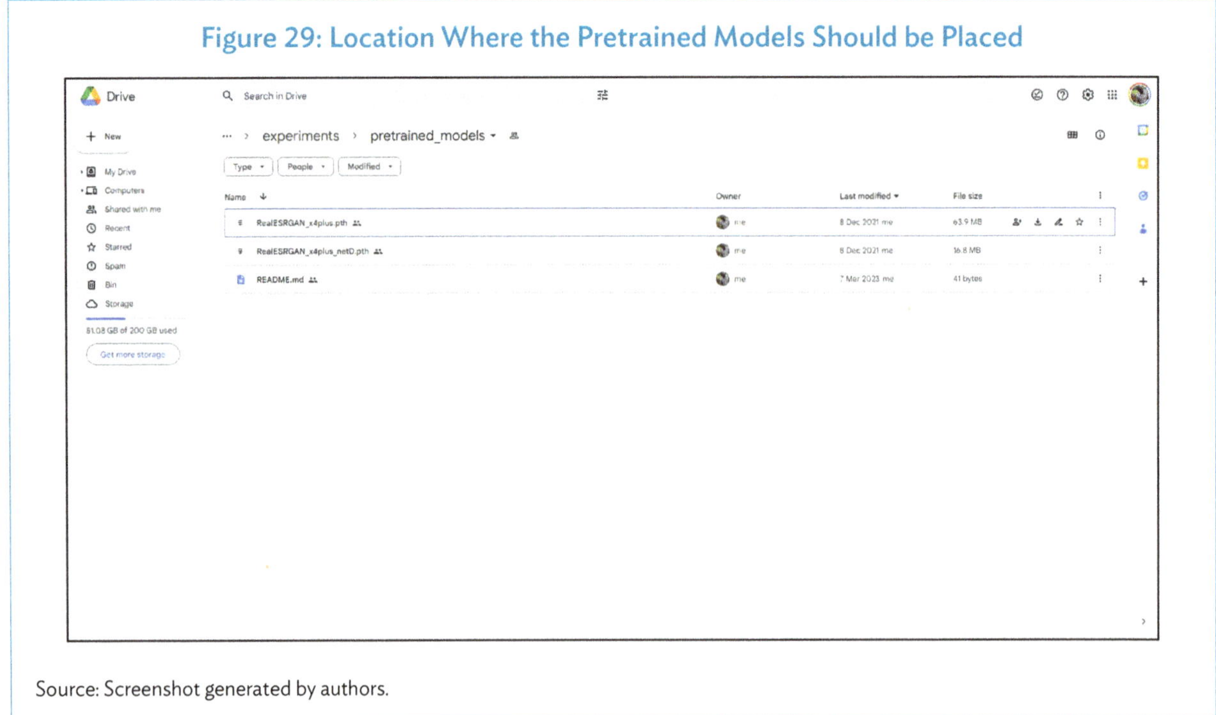

Source: Screenshot generated by authors.

Next, we run the following command to train the Generator model:

→ python realesrgan/train.py – opt options/finetune_realesrgan_roads.yml – auto resume

→ python realesrgan/train.py – opt options/finetune_realesrgan_roads.yml – auto resume

Code on Training the Generator Model

```
python realesrgan/train.py -opt options/finetune_realesrgan_roads.yml --
auto resume
```

The output generated appears as shown below:

Sample of Generated Output

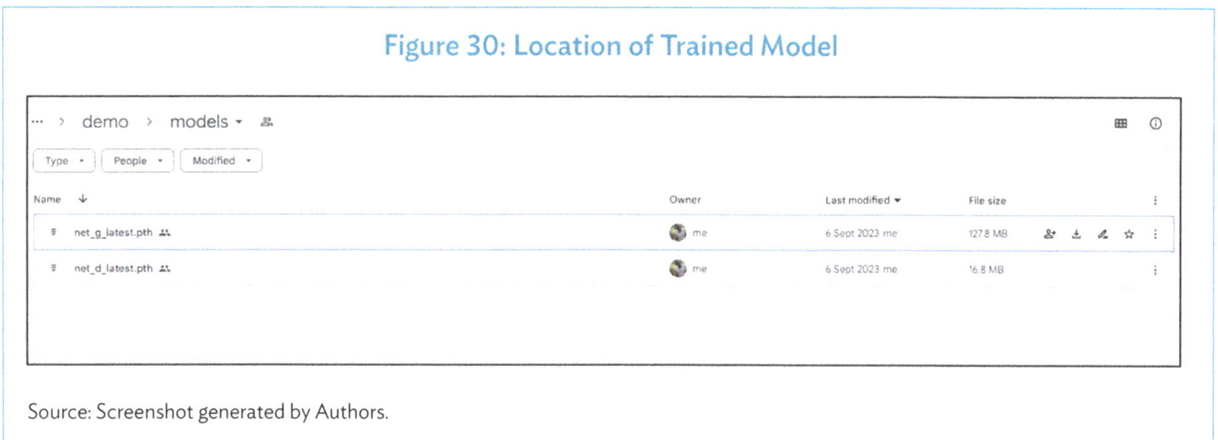

Once the training session is finished, you should have some pretrained models in your `experiments/your session name/models` folder:

Figure 30: Location of Trained Model

Name ↓	Owner	Last modified ▼	File size	
net_g_latest.pth	me	6 Sept 2023 me	127.8 MB	
net_d_latest.pth	me	6 Sept 2023 me	16.8 MB	

Source: Screenshot generated by Authors.

The net_g_latest.pth is what we use to super resolve our low-resolution images. A sample test of three images using the latest model provides the following results:

Figure 31: Sample of Super-Resolved Imageries

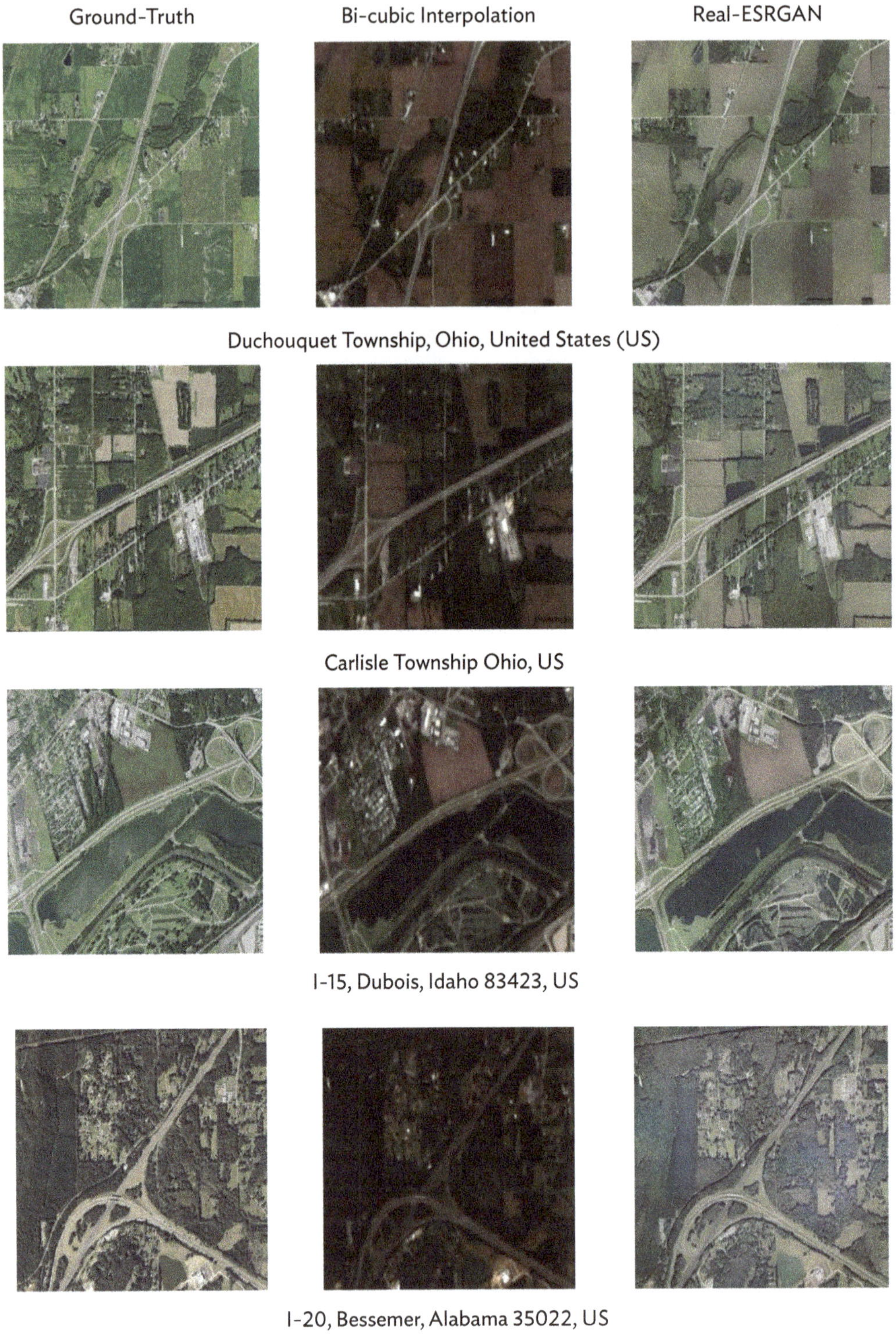

HR NAIP image (left column) of sample roads in the United States. HR image obtained from model output (right column) and interpolated image (middle column) obtained by bi-cubic interpolation.

Source: NAIP images downloaded from Google Earth Engine: https://developers.google.com/earth-engine/datasets/catalog/USDA_NAIP_DOQQ.

To use this pretrained model, we need to modify the "inference.py" to accommodate the model we trained. This can be found in the root of the Real-ESRGAN directory. One should also move the model file into the weight's directory, the root Real-ESRGAN directory:

Code Necessary to Use the Pretrained Model

```python
# determine models according to model names
args.model_name = args.model_name.split('.')[0]
if args.model_name == 'RealESRGAN_x4plus':  # x4 RRDBNet model
    model = RRDBNet(num_in_ch=3, num_out_ch=3, num_feat=64, num_block=23, num_grow_ch=32, scale=4)
    netscale = 4
    file_url = ['https://github.com/xinntao/Real-ESRGAN/releases/download/v0.1.1/RealESRNet_x4plus.pth']
elif args.model_name in ['netward_100k', 'RealEsrNet50k', 'net_g_latest']:# x4 RRDBNet model
    model = RRDBNet(num_in_ch=3, num_out_ch=3, num_feat=64, num_block=23, num_grow_ch=32, scale=4)
    netscale = 4

elif args.model_name == 'RealESRNet_x4plus':  # x4 RRDBNet model
    model = RRDBNet(num_in_ch=3, num_out_ch=3, num_feat=64, num_block=23, num_grow_ch=32, scale=4)
    netscale = 4
    file_url = ['https://github.com/xinntao/Real-ESRGAN/releases/download/v0.1.1/RealESRNet_x4plus.pth']
elif args.model_name == 'RealESRGAN_x4plus_anime_6B':  # x4 RRDBNet model with 6 blocks
    model = RRDBNet(num_in_ch=3, num_out_ch=3, num_feat=64, num_block=6, num_grow_ch=32, scale=4)
    netscale = 4
```

VII. Classification Using Satellite Imagery

The super-resolution model can now be used to improve the resolution of the imagery that has been downloaded around the centroids from the Philippines. Since this imagery hasn't been downloaded, we use the Colab notebook "**philippines_download_sections.ipynb**."

Downloading the Philippines' Road Sections

Step 1.

First begin by installing the following dependencies:

- → !pip install geemap
- → !pip install haversine

Geemap is used to download the satellite imagery and haversine is used to extract spheroidal distances between points relative to the earth.

Code on Installing Geemap and Haversine

Step 2.

The next step involves importing the necessary modules required in the downloading session and the initialization of both the Earth Engine API and the Google Drive API in order to store our imagery within the Google Drive directory.

- → import ee, os, pandas as pd, time, logging, geemap, glob as glob
- → from os.path import join
- → from multiprocessing import Pool
- → from haversine import haversine, Unit
- → ee.Authenticate()

→   ee.Initialize(opt_url='https://earthengine-highvolume.googleapis.com')

→   from google.colab import drive

→   drive.mount('/content/gdrive', force_remount=True)

Code on Importing Necessary Modules

```
[ ]  # utils
     import ee, os, pandas as pd, time, logging, geemap, glob as glob
     from os.path import join
     from multiprocessing import Pool
     from haversine import haversine, Unit


     ee.Authenticate()
     ee.Initialize(opt_url='https://earthengine-highvolume.googleapis.com')

     To authorize access needed by Earth Engine, open the following URL in a web browser and follow the instructions. I

        https://code.earthengine.google.com/client-auth?scopes=https%3A//www.googleapis.com/auth/earthengine%20https%3/

     The authorization workflow will generate a code, which you should paste in the box below.
     Enter verification code: 4/1Adeu5BVlAClak5ncrqVSguOQdd7u-ynPFqcyNxchbrF-hUw1ZDB1o0GpfcQ

     Successfully saved authorization token.

 ▶   from google.colab import drive
     drive.mount('/content/gdrive', force_remount=True)

 ⊡  Mounted at /content/gdrive
```

Step 3.

The next steps involve generating some utility directories that are used to download imagery into the predefined Google Drive directory.

To change the root directory:

→   root_dir = 'path to gdrive folder'

→   iri_sentinel_dir = 'path to gdrive folder'

To set the dates of the imageries to be downloaded:

→   *date_start = 'YYYY-MM-DD' (target start of imagery)*

→   *date_end = 'YYYY-MM-DD' (target end of imagery)*

To set the resolution of the imageries:

→   *lr_scale = set resolution image scale (for sentinel, 10 meters per pixel)*

To ensure that Python recognizes the directory:

→   *if not os.path.isdir(iri_sentinel_dir):*
        os.makedirs(iri_sentinel_dir, exist_ok=True)
→   *os.path.isdir(iri_sentinel_dir)*

Code on Generating Utility Directories

```
▾ Download Philippines IRI data from Sentinel satellites

[ ]  # change to root dir
     root_dir = '/content/gdrive/MyDrive/DEMO'
     iri_sentinel_dir =  '/content/gdrive/MyDrive/DEMO/phil_rd_sections_cliff'


     # Key variables
     date_start = '2019-06-01'
     date_end = '2019-12-30'

     # Set lower resolution image scale
     # For sentinel 10 meters per pixel
     lr_scale = 10


[ ]  if not os.path.isdir(iri_sentinel_dir):
         os.makedirs(iri_sentinel_dir, exist_ok=True)


[ ]  os.path.isdir(iri_sentinel_dir)

     True
```

Step 4.

We read our bbox CSV file that contains the latitude and longitude coordinates of the Philippines Road sections.

To read the international roughness index (IRI) coordinate information from CSV file:

→   df = pd.read_csv(join(root_dir, 'filename of CSV file'))
→   print(df.shape[0])
→   df.sort_values(by=["OBJECTID"], inplace = True)

To convert coordinate information data frame to list for processing:

→   *data_inputs = df.vaues.tolist()*
→   *print(data_inputs[0])*

Code to Read IRI Coordinate Information and Convert the Data Frame to List

```
[ ]  # Read IRI coordinate information from CSV file
     df = pd.read_csv(join(root_dir,'bounding_boxes_mercator_date.csv'))
     print(df.shape[0])
     df.sort_values(by=["OBJECTID"], inplace = True)
     # Convert coordinate information dataframe to list for processing
     data_inputs = df.values.tolist()
     print(data_inputs[0])

     124276
     [4726, 'S00001BR', 1, 11.45391442, 124.4784091, 11.45474062, 124.47872, 2019, 3, 1]
```

Note: First sort the data frame in ascending order and then convert the pandas data frame into a list object.

Step 5.

Next is to identify and mask out cloudy and cirrus pixels in S2 satellite images. This will ensure that the processed images are free from cloud cover interference. The maskS2clouds function is designed for this step.

→   def maskS2clouds(image):
         qa = image.select('QA60');
→   cloudBitMask = 1 << 10;
→   cirrusBitMask = 1 << 11;
→   mask = qa.bitwiseAnd(cloudBitMask).eq(0)
→   mask = mask.And(mask.bitwiseAnd(cirrusBitMask).eq(0))
→   return image.updateMask(mask).divide(10000)

Code on Identifying and Masking Out Cloudy and Cirrus Pixels

```
[ ]  # Function to mask clouds
     def maskS2clouds(image):
       qa = image.select('QA60');

       #Bits 10 and 11 are clouds and cirrus, respectively.
       cloudBitMask = 1 << 10;
       cirrusBitMask = 1 << 11;

       #Both flags should be set to zero, indicating clear conditions.
       mask = qa.bitwiseAnd(cloudBitMask).eq(0)
       mask = mask.And(mask.bitwiseAnd(cirrusBitMask).eq(0))

       return image.updateMask(mask).divide(10000)
```

Step 6.

The next step involves downloading the Philippines satellite road sections using the function "download satellite sections." It does the following:

- **Bounding box extraction:** It extracts a bbox from the input road_sec, which is a list of values containing information about a road section. The bbox is defined by the coordinates in road_sec and is used to specify the region of interest (ROI) for image retrieval.
- **Buffering the region:** It creates a buffer around the bounding box (AOI) with a specified buffer distance (buffer). This buffer helps in including additional area around the road section in the ROI.
- **Calculating dimensions:** It calculates the dimensions (width and height) of the ROI in pixels based on the difference in coordinates and a scaling factor (lr_scale). The haversine function seems to be used to calculate distances between coordinates.
- **Constructing image collection:** It constructs an Earth Engine image collection (sentinel) by filtering Sentinel-2 satellite images based on the ROI, date range (date_start to date_end), and cloud cover percentage criteria. It also applies a cloud mask (maskS2clouds) to filter out cloudy pixels.
- **Downloading image:** If there are images in the sentinel collection, it calculates visual parameters (vis_params) for image visualization, specifies the image filename (x), and uses the Geemap library to fetch the image thumbnail from Google Earth Engine (GEE). The downloaded image is saved as a JPEG file.
- **Main execution:** The function is set to execute when the script is run directly (if __name__ == '__main__':). It uses a multiprocessing pool (Pool) to parallelize the download process for a list of road sections (data_inputs) with a pool size of 25 processes. This allows multiple images to be downloaded concurrently, which can significantly speed up the process.

For this step, you may refer to the code below:

```
→   def download_road_images2(road_sec):
        bbox = [your bounding boxes containing coordinate]
        buffer = 500
        region ee.Geometry.Rectangle(bbox)
        aoi = region.buffer(buffer).bounds()
        long_start = aoigetInfo()["coordinates"][0][0][0]
        long_end = aoi.getInfo()["coordinates"][0][1][0]
        lat_start = aoi.getInfo()["coordinates"][0][0][1]
        lat_end = aoi.getInfo()["coordinates"][0][2][1]

        a = (lat_start, long_start)
        b = (lat_start, long_end)
        width = haversine(a,b)
        px_width = int(width*1000/lr_scale)
        a = (lat_start, long_start)
```

```
b = (lat_end, long_start)
height = haversine(a,b)
px_height = int(height*1000/lr_scale)

dim = str(px_width)+'x'+str(px_height)
sentinel = (ee.ImageCollection('COPERNICUS/S2_SR_HARMONIZED')
    .filterBounds(aoi)
    .filterDate(date_start, date_end)
    .filter(ee.Filter.lt('CLOUDY_PIXEL_PERCENTAGE', 60)).map(maskS2clouds))
If sentinel.size().getInfo() > 0:
    Image = sentinel.median().clip(aoi)
    Vis_params = {'min': 0.0, 'max': 0.3, 'bands': ['B4', 'B3', 'B2'], }
```

Code on Downloading the Philippines Satellite Road Sections

```python
def download_road_images2(road_sec):
    bbox = [road_sec[4], road_sec[3], road_sec[6], road_sec[5]]#your bounding box containing coordinates

    buffer = 500
    #Geometry being a rectangle
    region = ee.Geometry.Rectangle(bbox)
    # buffer your rectangele with bounds
    aoi = region.buffer(buffer).bounds()

    long_start = aoi.getInfo()["coordinates"][0][0][0]
    long_end = aoi.getInfo()["coordinates"][0][1][0]
    lat_start = aoi.getInfo()["coordinates"][0][0][1]
    lat_end = aoi.getInfo()["coordinates"][0][2][1]
    # print(lat_start, long_start)
    # Width (diff between longitudes)
    a = (lat_start, long_start)
    b = (lat_start, long_end)
    width = haversine(a, b)
    px_width = int(width*1000/lr_scale)

    # Height (diff between latitudes)
    a = (lat_start, long_start)
    a = (lat_end, long_start)
    height = haversine(a,b)
    px_height = int(height*1000/lr_scale)

    dim = str(px_width)+'x'+str(px_height)
    # print(dim)
    sentinel = (ee.ImageCollection('COPERNICUS/S2_SR_HARMONIZED')
            .filterBounds(aoi)
            .filterDate(date_start,date_end)
            .filter(ee.Filter.lt('CLOUDY_PIXEL_PERCENTAGE', 60)).map(maskS2clouds)
    )

    if sentinel.size().getInfo() > 0:
        image = sentinel.median().clip(aoi)
        vis_params = {
                'min': 0.0,
                'max': 0.3,
                'bands': ['B4', 'B3', 'B2'],
                }
```

Step 7.

We need to get a list of downloaded filenames and map them to the IRI data CSV file, which involves first getting the filenames and mapping them into a data frame:

```
→    import glob as glob
→    vis = []
→    for file in os.listdir(iri_sentinel_dir):
         if file.endswith(".jpg"):
             vis +=[file]
→    dictoid = {'filenames': vis}
→    df_train = pd.DataFrame(dictoid)
→    print('Number of image segments: '+str(len(df_train)))
→    object_ids = []
→    fnames = []
→    for d in df_train.itertuples():
         f = d.filenames
         x = f.split('.')
         fnames += [f]
         object_ids += [int(x[0])]
→    dict_seg = {'filenames': fnames, 'OBJECTID': object_ids}
→    df_SEG = pd.DataFrame(dict_seg)
```

Then to merge the filename Data Frame with the IRI CSV file and dropping the unnecessary columns:

```
→    df_IRI = pd.read_csv('IRI_data.csv', sep=',', encoding='latin-1')

→    df_merge = pd.merge(left = df_SEG, right = pd.df_IRI, on = ['OBJECTID'], how = 'left')

→    df_merge.drop(['SECTION_ID', 'AVE_IRI', 'ISLAND', 'REGION', 'PROVINCE', 'DEO',
     'ROAD_NAME', 'ROAD_CLASS', 'ROUTE_NO', 'CONG_DIST'])
→    path = os.path.join(root_dir, 'filename of merged file')
→    with open(path, 'w', encoding = 'utf-8-sig') as f:
         df_merge.to_csv(f)
→    pd.read_csv(path).head(10)
```

Code on Getting Filenames and Mapping Them into a Data Frame

```python
# Store list of filenames in directory within a list object
import glob as glob
vis = []

for file in os.listdir(iri_sentinel_dir):
    if file.endswith(".jpg"):
        vis +=[file]

dictoid = {'filenames': vis}
df_train = pd.DataFrame(dictoid)

print('Number of image segments: '+str(len(df_train)))

# Extact object IDs from filenames
object_ids = []
fnames = []

for d in df_train.itertuples():
    f = d.filenames
    x = f.split('.')
    fnames += [f]
    object_ids += [int(x[0])]

dict_seg = {'filenames': fnames, 'OBJECTID': object_ids}
df_SEG = pd.DataFrame(dict_seg)
```

```python
# Merge with IRI csv file

# Read the iri csv file
df_IRI = pd.read_csv('IRI_data.csv',  sep=',', encoding='latin-1')

# Merge
df_merge = pd.merge(left = df_SEG, right = df_IRI, on = ['OBJECTID'], how = 'left')

# Drop non essential columns.
# All that is required is filename and corresponding IRI label.
df_merge.drop(['SECTION_ID', 'AVE_IRI','ISLAND', 'REGION', 'PROVINCE', 'DEO', 'ROAD_NAME','ROAD_CLASS', 'ROUTE_NO','CONG_DIST',

# Save the merged file to a .csv
path = os.path.join(root_dir ,'demo_output_cliff.csv')

with open(path, 'w', encoding = 'utf-8-sig') as f:
  df_merge.to_csv(f)

pd.read_csv(path).head(10)
```

We now have training data that we can feed into the convolutional neural network (CNN).

Training the Classification Model

Step 1

Begin by importing the necessary modules:

- → import numpy as np
- → import pandas as pd
- → from PIL import Image
- → import torch
- → from tqdm.auto import tqdm
- → import torchvision
- → import shutil
- → import matplotlib.pyplot as plt
- → from torch.utils.data import Dataset
- → from sklearn.model_selection import train_test_split
- → from torch.utils.data import Dataloader
- → from torch import nn, optim
- → from torch.nn import Sequential
- → import torch.nn.functional as F
- → from torchvision import models, transforms
- → from torchvision.models import ResNet34_Weights, EfficientNet_B4_Weights, EfficientNet_B0_Weights
- → from torchvision.transforms import CenterCrop, Resize
- → from sklearn.preprocessing import LabelEncoder

Code on Importing Necessary Modules

```
Modules

# namespaces
import numpy as np
import pandas as pd
from PIL import Image

import torch
from tqdm.auto import tqdm
import torchvision
import shutil
import matplotlib.pyplot as plt
from torch.utils.data import Dataset
from sklearn.model_selection import train_test_split
from torch.utils.data import DataLoader
from torch import nn, optim
from torch.nn import Sequential
import torch.nn.functional as F
from torchvision import models, transforms
from torchvision.models import ResNet34_Weights, EfficientNet_B4_Weights, EfficientNet_B0_Weights
from torchvision.transforms import CenterCrop, Resize
from sklearn.preprocessing import LabelEncoder
```

Step 2

Keep in mind that this is the section where we can do a four-class or two-class image classification. If we are doing a two-class image classification, we convert "fair" into "good" and "poor" into "bad".

Load in the CSV we generated for use with the images that we downloaded to carry out classification:

→   root = 'path to your gdrive'

→   image_path = os.path.join(root, "path to image folder")

→   csv_target_path = os.path.join(root, "path to csv file")

→   df = pd.read_csv(csv_target_path)

→   df.drop(["Unnamed: 0", "filenames"], inplace=True, axis=1)

→   existing_files = os.listdir(image_path)

→   df["file_name"] = df.apply(lambda x: f"{{}}.jpg".format(x["OBJECTIS"]), axis=1)

→   df = df[df["file_name"].isin(existing_files)]

Code on Loading in the CSV Files

```
DEMO CSV

▶   root = '/content/gdrive/MyDrive'
    image_path = os.path.join(root, "DEMO/phil_rd_sections_cliff")
    csv_target_path = os.path.join(root, "DEMO/demo_output_cliff.csv")

    df = pd.read_csv(csv_target_path)
    df.drop(["Unnamed: 0", "filenames"], inplace=True, axis=1)


[ ]  existing_files = os.listdir(image_path)

    df["file_name"] = df.apply(lambda x: f"{{}}.jpg".format(x["OBJECTID"]), axis=1)
    df = df[df["file_name"].isin(existing_files)]
```

Step 3

The next section prepares a custom dataset for training a machine learning model using PyTorch:

1.   Label Encoding:
     – le = LabelEncoder(): This creates an instance of the LabelEncoder class, which is used for encoding categorical labels (in this case, IRI_RATING) into numerical values.
     – label_encoded = le.fit_transform(df['IRI_RATING']): It applies label encoding to the 'IRI_RATING' column of the DataFrame df. The encoded labels are stored in the label_encoded variable.
     – df['label_encoded'] = label_encoded: The encoded labels are added as a new column 'label_encoded' to the DataFrame df. This column will be used as target labels for the machine learning model.

2. **Train-Validation Split:**
    - train_df , val_df = train_test_split(df, test_size=0.2, stratify=df.IRI_RATING, random_
      state=42): This code splits the DataFrame df into training (train_df) and validation (val_df)
      subsets. It uses train_test_split from sci-kit-learn to do the split with an 80-20 ratio.
    - stratify=df.IRI_RATING ensures that the class distribution in the original data is preserved in
      both the training and validation sets.

3. **Custom Dataset Definition:**
    - The class CustomDatasetFromImages(Dataset): Defines a custom dataset class that
      inherits from PyTorch's Dataset class.
    - In the constructor (__init__ method), the dataset is initialized with a DataFrame
      (dataframe), a directory containing images (img_dir), and an optional transformation
      (transform) for data augmentation. The constructor reads the DataFrame, extracts image
      paths and labels, and calculates the length of the dataset.
    - The __getitem__ method is responsible for loading and preprocessing individual data
      samples. It loads an image, applies the specified transformations, and returns the image
      tensor and its corresponding label.
    - The get_labels method returns the labels of the dataset.
    - The __len__ method returns the total number of samples in the dataset.

For this step, you may refer to the code below:

```
→   Le = LabelEncoder()
→   label_encoded = le.fit_transform(df['IRI_RATING'])
→   df['label_encoded'] = label_encoded
→   class CustomDatasetFromImages(Dataset):
        def __init__(self, dataframe, img_dir, transform = None):
            self.to_tensor = transforms.ToTensor()
            self.transform = transform
            self.data_info = dataframe
            self.image_arr = np.asarray(self.data_info.iloc[:, 2])
            self.label_arr = np.asarray(self.data_info.iloc[:, 3])
            self.img_dir = img_dir
            self.data_len = len(self.data_info.index)
        def __getitem__(self, index):
            single_image_name = self.image_arr[index]
            img_path = os.path.join(self.img_dir, single_image_name)
```

Code on Preparing Custom Datasets

```
[ ]  #Label Encoder

     le = LabelEncoder()
     label_encoded = le.fit_transform(df['IRI_RATING'])
     df['label_encoded'] = label_encoded

     train_df , val_df = train_test_split(df, test_size = 0.2, stratify = df.IRI_RATING, random_state = 42)
     #--------------------------------Class Data --------------------------------------------------------#
     class CustomDatasetFromImages(Dataset):
         def __init__(self, dataframe, img_dir,  transform = None):
             """

             Args:
                 dataframe (string): csv file
                 img_path (string): path to the folder where images are
                 transform: pytorch transforms for transforms and tensor conversion
             """

             # Transforms
             self.to_tensor = transforms.ToTensor()
             self.transform = transform
             # Read the csv file
             self.data_info = dataframe
             # First column contains the image paths
             # must be an np.array type to avoid memory explosion bug from torch dataloader and numworker iter
             self.image_arr = np.asarray(self.data_info.iloc[:, 2])
             # Second column is the labels
             self.label_arr = np.asarray(self.data_info.iloc[:, 3])

             self.img_dir = img_dir
             # Calculate len
             self.data_len = len(self.data_info.index)

         def __getitem__(self, index):
             # Get image name from the pandas df
             single_image_name = self.image_arr[index]
             # Open image
             img_path = os.path.join(self.img_dir , single_image_name)
```

Step 4

Keep in mind that this is the section where one can choose whether to use resnet34 or efficientnetb0.

The next section sets up data augmentation transformations for different model types and augmentation levels. We then initialize training and validation datasets (train_data and val_data) using a custom dataset class (CustomDatasetFromImages) and apply the specified augmentations to the images based on the augmentation's variable, which is determined by the combination of model_type and augment values.

→   elif augment == 2:
         augmentations = Sequential(Resize([384,384]), CenterCrop([380,380]),)
→   elif augment == 3:
         augmentations = Resize([380,380])
→   else:
         augmentations = None

```
→    elif model_type =="efficientnet_b0:
        if augment == 1:
            augmentations = CenterCrop([224,224])
        elif augment == 2:
            augmentations = Resize ([895,896])
        elif augment == 3:
            augmentations = Resize ([224,224])
        else:
            augmentations = None
→    train_data = CustomDatasetFromImages(train_df, image_path, transform = augmentations)
→    val_data = CustomDatasetFromImages(val_df, image_path, transform = augmentations)
```

Code on Setting Up Data Augmentation Transformations

```
            ])
  elif augment == 2:
    augmentations = Sequential(
                Resize([384,384]),
                CenterCrop([380,380]),
                )


  elif augment == 3:
    augmentations = Resize([380,380])


  else:
    augmentations = None

elif model_type == "efficientnet_b0":
  if augment == 1:
    augmentations = CenterCrop([224,224])
  elif augment == 2:
    augmentations = Resize([895,896])
  elif augment == 3:
    augmentations = Resize([224,224])
  else:
    augmentations = None


train_data = CustomDatasetFromImages(train_df, image_path, transform = augmentations)
val_data = CustomDatasetFromImages(val_df, image_path, transform = augmentations)
```

Step 5

We then define a function for training our model using the data that we downloaded. Inside the training loop, which iterates through batches of training data:

optimizer.zero_grad(set_to_none=True) clears the gradients of all optimized variables, preparing them for a new gradient calculation for the current batch.

inputs = inputs.to(device, non_blocking=True) moves the input data (typically images) to the specified computing device (e.g., GPU) for accelerated computation.

labels = labels.to(device, non_blocking=True) moves the corresponding labels to the same computing device.

prediction = model(image) performs a forward pass of the model with the input data to generate predictions.

loss = criterion(prediction, labels) calculates the loss between the model's predictions and the actual labels using a specified loss criterion (e.g., cross-entropy loss).

loss.backward() computes the gradients of the loss with respect to the model's parameters, enabling backpropagation.

optimizer.step() updates the model's parameters using the computed gradients and the chosen optimization algorithm (e.g., stochastic gradient descent).

tloss += loss.item() is the training loss for the current batch is accumulated in tloss.

For this step, you may refer to the code below:

```
→   def train_model(starting_epoch, num_epochs, model, train_dataloader, val_dataloader,
        training_losses: list, valid_losses: list, save_checkpoint, path):
        for epoch in range(starting_epoch, num_epochs):
            tloss = 0
            vloss = 0
        running_accuracy = 0.0
        total = 0
        with tqdm(train_dataloader, unit= 'batch') as train_loader:
            model.train(True)
            for i, (inputs, labels) in enumerate(train_loader):
            optimizer.zero_grad(set_to_none=True)
            inputs = inputs.to(device, non_blocking=True)
            labels = labels.to(device, non_blocking=True)
            image=inputs.float()
            prediction = model(image)
            loss = criterion(prediction,labels)
```

```
loss.tebackward()
optimizer.step()
tloss += loss.item()
train_loader.set_postfix(stats='Epoch: {} \tTraining Loss: {:.6f}'.format(epoch, tloss/
len(train_loader))))
epoch_loss = tloss/len(train_dataloader)
training_losses.append(epoch_loss)
torch.cuda.empty_cache()
with tqdm(val_dataloader, unit="batch") as valid_loader:
    model.train(False)
    for i, (inputs,labels) in enumerate(valid_loader):
    with torch.no_grad():
    inputs = inputs.to(device, non_blocking=True)
    labels = labels.to(device, non_blocking=True)
    image=inputs.float()
```

Code on Defining a Function for Training the Model

```python
def train_model(starting_epoch, num_epochs, model, train_dataloader, val_dataloader, training_losses: list, valid_losses: list, save_checkpoint, path):
    # inner variables
    for epoch in range(starting_epoch, num_epochs):
        tloss = 0
        vloss = 0
        running_accuracy = 0.0
        total = 0
        with tqdm(train_dataloader, unit= 'batch') as train_loader:
            model.train(True) #set the model to training mode
            for i, (inputs, labels) in enumerate(train_loader):
                # clear the gradients of all optimized variables
                optimizer.zero_grad(set_to_none=True)
                inputs = inputs.to(device, non_blocking=True)
                labels = labels.to(device, non_blocking=True)
                image=inputs.float()
                # forward pass: compute predicted outputs by passing inputs to the model
                prediction = model(image)
                # calculate the batch loss
                loss = criterion(prediction,labels)
                # backward pass: compute gradient of the loss with respect to model parameters
                loss.backward()
                # perform a single optimization step (parameter update)
                optimizer.step()
                #log training loss
                tloss += loss.item()
                train_loader.set_postfix(stats='Epoch: {} \tTraining Loss: {:.6f}'.format(epoch, tloss/len(train_loader)))
        epoch_loss = tloss/len(train_dataloader)
        training_losses.append(epoch_loss)
        torch.cuda.empty_cache()
        ########################
        # validate the model #
        ########################
        with tqdm(val_dataloader, unit="batch") as valid_loader:
            model.train(False)
            for i, (inputs, labels) in enumerate(valid_loader):
                with torch.no_grad():
                    inputs = inputs.to(device, non_blocking=True)
                    labels = labels.to(device, non_blocking=True)
                    image=inputs.float()
```

We can now initialize the model in the main method, reshape it, specify our gradient descent and loss function then pass a list of keyword arguments to begin the training:

→   number of epochs = 10

→   train_model(0, number_of_epochs, model, train_dataloader, val_dataloader, training_losses, valid_losses, True, model_dir)

Code on Initializing the Model in the Main Method

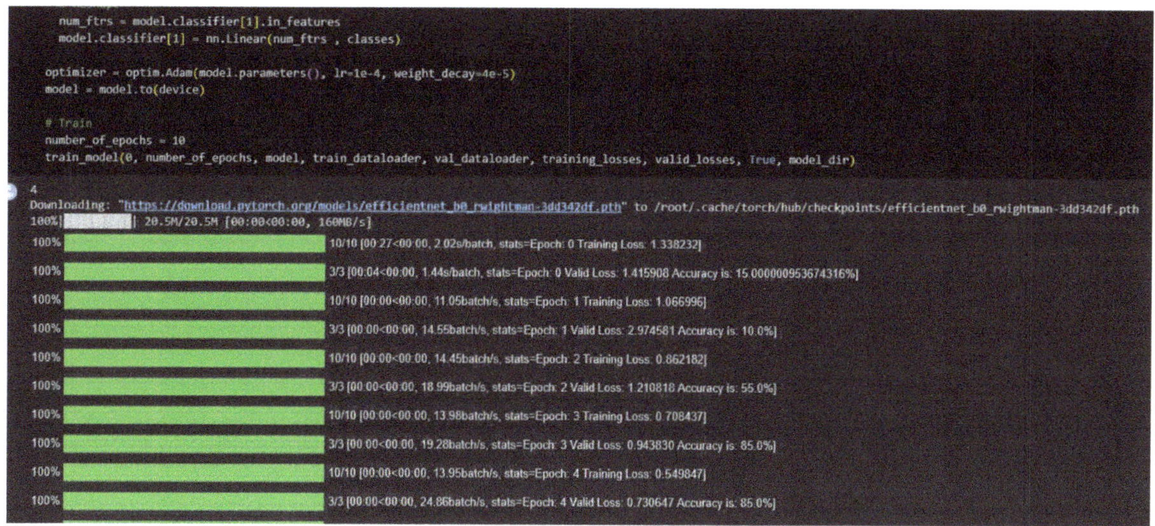

We can then evaluate our model by running predictions on the validation sets and plotting a confusion matrix.

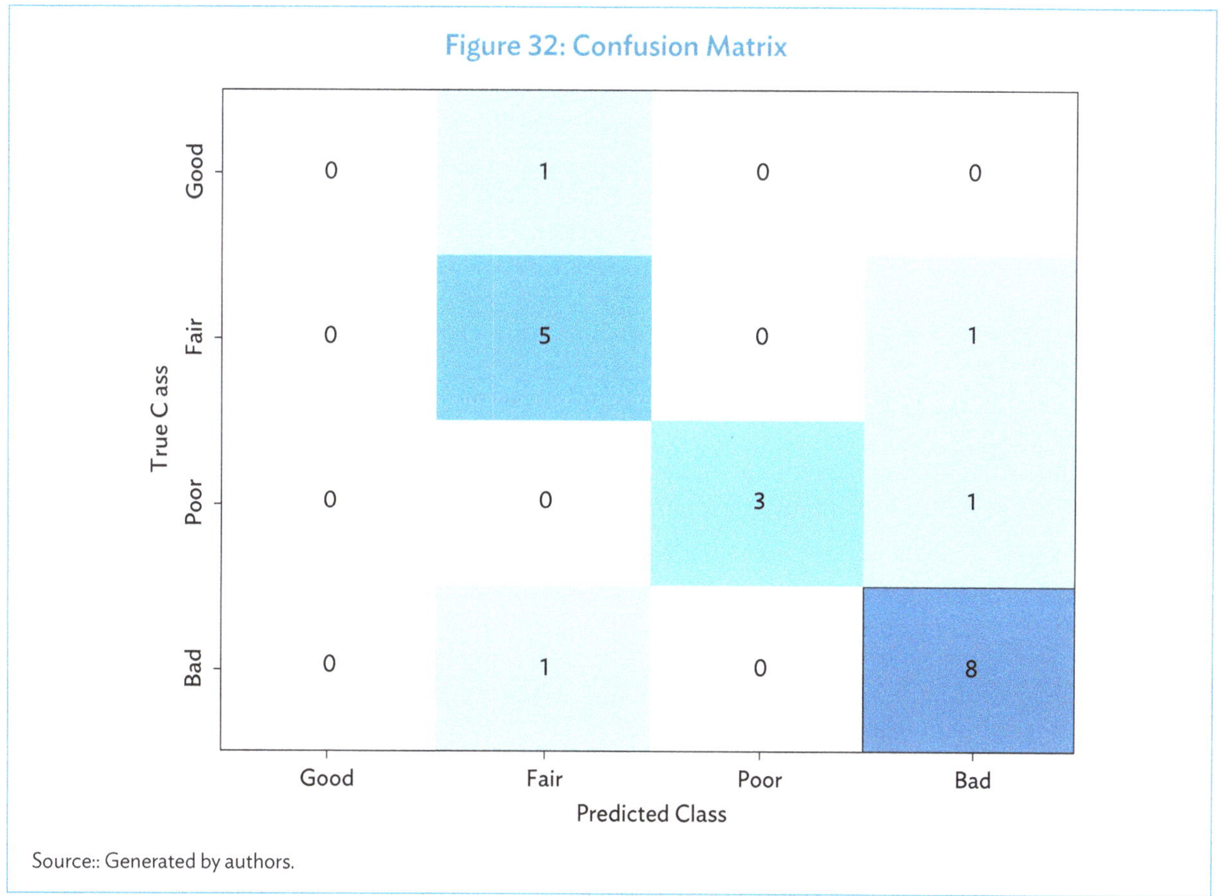

Figure 32: Confusion Matrix

Source:: Generated by authors.

VIII. Alternative Method— Smartphone-Based Pavement Condition Assessment

The popularity of the smartphone, the improved technology of the device, as well as its better access to the internet and global positioning system (GPS) connectivity in most geographic locations of a country have made the use of smartphone data to monitor pavement conditions a viable option for most road agencies.

Smartphone-based pavement condition assessment primarily focuses on ride quality evaluation. The vibrations caused by the unevenness of the pavement surface or distress affect the ride quality. This is evaluated using metrics such as pavement serviceability rating, international roughness index, or suitable pavement condition evaluation indices. Moreover, the image data can be also used to assess the overall pavement condition utilizing image processing techniques. The following section discusses these aspects.

Smartphone Technology Used for Pavement Condition Assessment

Most new smartphones have in-built three-axis accelerometers and GPS capabilities. The measurement of road roughness utilizes the accelerometer to measure vertical vehicle-body acceleration. The presence of defects, such as potholes, can be recognized by identifying relatively large variations in continuous acceleration recordings. Road roughness can be determined by transforming the continuous vertical acceleration recordings and matching these with the GPS-determined locations.

Acceleration data, GPS data, time, and image/video data from mobile cameras are the primary data sources used to evaluate pavement conditions using smartphones. There are various sensors embedded in smartphones that can be used either directly or indirectly to measure some of these parameters (Table 1).[17]

---

[17]   S. Sattar, S. Li, and M. Champmon. 2018. Road Surface Monitoring Using Smartphone Sensors: A Review. *Sensors.*

Table 1: Smartphone Sensors Used to Evaluate Pavement Condition Metrics

Name of the Sensor	Unit	Application in Pavement Condition Evaluation
Accelerometer	m/s^2	Measures the acceleration force
Gyroscope	rad/s	Measures a device's rate of rotation
Magnetometer	T	Measures the ambient geomagnetic field
Rotation	rad	Measures the orientation of a device
GPS	degree	Obtain location information, use for speed calculation
Camera	pixels	Obtain video/photos

GPS = global positioning system, m/s^2 = meter per second squared, rad/s = radian per second, T = tesla

Source: S. Sattar, S. Li, and M. Champmon. 2018. Road Surface Monitoring Using Smartphone Sensors: A Review. Sensors.

The accelerometer measures the acceleration in each direction (x, y, and z planes) with respect to the gravitational force, and it has specific patterns when the road condition varies. In addition to the distress detection, a threshold based on the average acceleration on road pavement would give an initial indication of the pavement quality level (Figure 33).

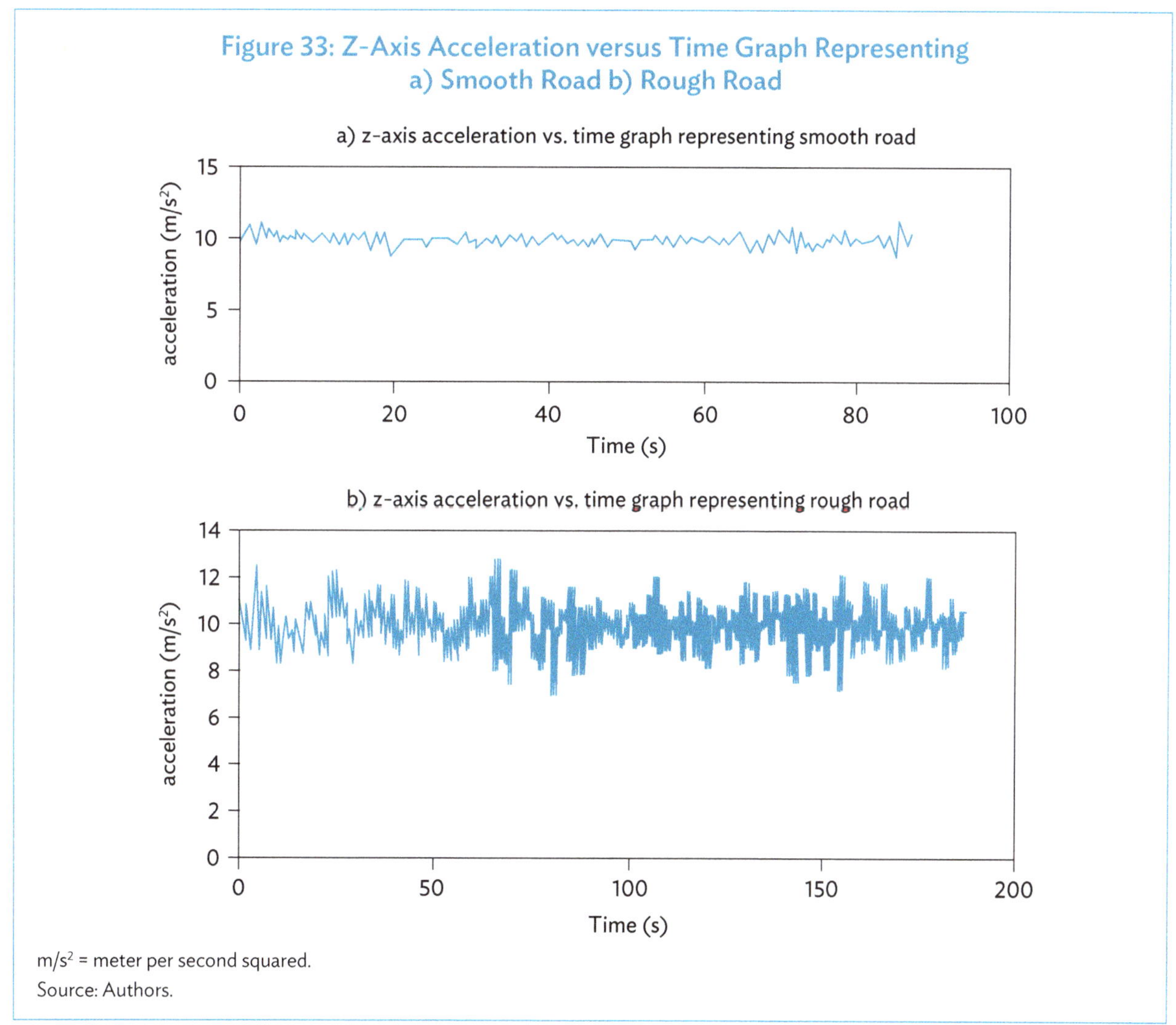

Figure 33: Z-Axis Acceleration versus Time Graph Representing a) Smooth Road b) Rough Road

m/s^2 = meter per second squared.

Source: Authors.

Various studies have shown that a number of characteristics of a smartphone may affect the measurement of vertical acceleration and therefore road roughness. The characteristics associated with the operating range, resolution, frequency, and sensitivity to gravity and temperature affect the accuracy and precision of the measurements. Moreover, the smartphone type and the position also have an impact on the accuracy of the acceleration data. An unstable position and different orientations of the smartphone would produce erroneous results. Therefore, to omit the error, the smartphone should be mounted on the windshield with a suction cup and with a bracket that attaches to the dashboard to ensure rigid mounting.

Several vehicular factors that can be affected in estimating smartphone roughness have been declared by the American Society for Testing Materials (ASTM) in 2012: vehicle speed, vehicle type, suspension stiffness and damping, tire pressure, sprung mass (or number of passengers/cargo), and acceleration/braking.[18] Increasing vehicle speed led to increases in simulated root-mean-square acceleration (Grms). Therefore, the survey speed should be kept between 20–100 kilometers per hour (km/h), and the IRI should calibrate for different speeds. Also, the tire pressure, sprung mass, and other vehicular factors shall be kept within the standard limits. The driving cycle also has a significant impact on smartphone roughness, hence the survey speed should be kept as constant as possible by not applying sudden braking/acceleration. Table 2 shows the impact level of each factor on the smartphone roughness measurement.[19]

Table 2: Summary of Different Influencing Factors on Simulated Vehicle Body Acceleration

Factor	Severity	Influence
Smartphone types	H	Very different in measuring IRI value but can be minimal in categorical analysis
Vehicle speeds	H	Simulated Grms increases logarithmically with speed. Increase of 93% in Grms with a speed increase of 266% (i.e., from 30–80 km/h)
Vehicle types	H	Differences in Grms between vehicle types of up to 190%
Sprung mass	M	Every 70 kg (i.e., approximately one additional passenger) added to the sprung mass results in approximately a 5% decrease in Grms
Longitudinal profile	M	Up to 25% change in Grms due to change in longitudinal profile
Suspension stiffness	M	5%–15% change in simulated Grms due to ±20% of changes in spring rate
Tire pressure	M	Up to 20% change in simulated Grms due to ±30% change in tire pressure
Driving style	L	3%–7% change in simulated Grms due to changing driving behavior
Suspension damping	L	Less than 8% change in simulated Grms as a result of ±20% of changes in damping

Grms = root-mean-square acceleration; H=High; IRI = international roughness index; kg = kilograms; km/h = kilometers per hour; L=Low; M=Moderate.

Source: G. Wang, M. Burrow, and G. Ghataora. 2020. Study of the Factors Affecting Road Roughness Measurement Using Smartphones. Journal of Infrastructure Systems

---

[18] ASTM. 2012. Standard test method for measurement vehicular response to traveled surface roughness. *ASTM E1082–90.*
[19] G. Wang, M. Burrow, and G. Ghataora. 2020. Study of the Factors Affecting Road Roughness Measurement Using Smartphones. *Journal of Infrastructure Systems.*

Since the IRI is estimated based on derived acceleration data of the smartphone, it is not a direct measurement of the surface profile, hence it falls under Class III response type roughness measurements. The analysis is often complex and has several limitations with respect to accuracy and precision. This is partly because a large number of factors need to be taken into account when determining road roughness from acceleration readings determined from a smartphone. As explained earlier, these factors are associated with the smartphones themselves, vehicle-related aspects, and the form of the road surface profile.

Roughness Evaluation Using Smartphone-Based Data

In most of smartphones, three-axis accelerometer is a common sensor which gives the acceleration along x, y, and z axis. The acceleration data can be used to derive the required parameters in the analytical approaches discussed below.

Vertical Displacement-Based Approach

Sayers and Karamihas[20] developed the quarter car model which can be used to compute the vertical movement of the vehicle body which produced variations due to road roughness-induced fluctuation. The basic approach in smartphone-based roughness measurement is converting the vehicle acceleration profile into the roughness profile by inverse calculation of vertical displacement. The IRI is estimated by back calculation using the integration equation as presented in Equation 1.

$$IRI = \int_0^{L/v} \frac{V_1 - V_2}{L} dt \qquad (1)$$

Where, V1 & V2 are velocity of the sprung and un-sprung mass, L is the length of the measure, V is the mean velocity of the vehicle.

In the equation, the sprung mass is the car body which is supported by the wheel, and un-sprung mass is the mass of wheel, tire & axle. The input required to estimate roughness by using the quarter car model is the wheel track elevation as a function of time. The output is produced as displacement, vertical velocity and acceleration.

Power Spectral Density (PSD) Approach

PSD is the measure of signal's power content versus frequency with the amplitude unit of g^2/Hz. A PSD is typically used to characterize broadband random signals and the amplitude is normalized by the spectral resolution employed to digitize the signal. The response of the Quarter car simulation model can also be deployed to determine the PSD of pavement roughness, and from this, the expression for IRI can be derived. Research has shown that the PSD of a vehicle's vertical acceleration, obtained from frequency domain analysis, correlates closely to road roughness[21]. This approach also follows

---

20   M. Sayers and S. Karamihas. 1998. *The Little Book of Profiling. Federal Highway Administration.*

21   L. Sun. 2001. Developing spectrum-based models for international roughness index and present serviceability index. *Journal of Transportation Engineering, 463–470*; J. A. Marcondes, M. Snyder, and P. Singh. 1992. Predicting vertical acceleration in vehicles through road roughness. *Journal of Transportation Engineering, 33–49*; R. Hesami and K. McManus. 2009. Signal processing approach to road roughness analysis and measurement. TENCON 2009–2009, IEEE Region 10 Conf. Singapore: IEEE.

the quarter car model principle to derive IRI value. The model follows a generic form as shown in Equation 2.

$$IRI = A\sqrt{PSD}$$

(2)

Where A is the constant and PSD is the power spectral density measured in g^2/Hz.

The model coefficients can be calibrated by regression analysis using the IRI estimated from the existing Class I or Class II type roughness measuring equipment and the square root of the PSD function which is calculated from the smartphone accelerometer readings. A study done by Janani, Sunitha, and Mathew[22], derived a relationship following this approach, which used roughness measurements from the device called Roughometer to estimate the coefficients which are given in Equation 3 and Figure 34.[22] The equation was derived after filtering the effects of major distresses such as potholes and it gives a good correlation between the established roughness measurement devices over a reasonable roughness value range.

$$IRI = 0.774\sqrt{PSD} - 0.825$$

(3)

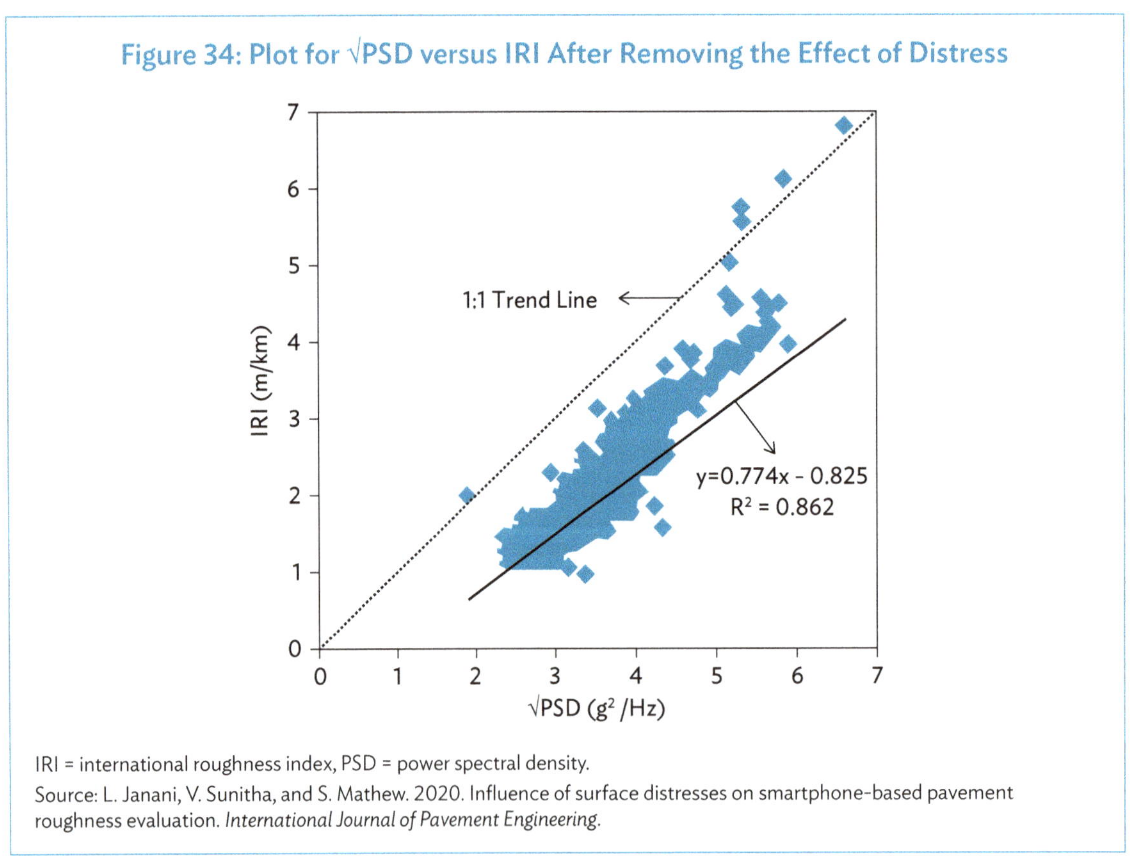

Figure 34: Plot for √PSD versus IRI After Removing the Effect of Distress

IRI = international roughness index, PSD = power spectral density.
Source: L. Janani, V. Sunitha, and S. Mathew. 2020. Influence of surface distresses on smartphone-based pavement roughness evaluation. *International Journal of Pavement Engineering*.

22   L. Janani, V. Sunitha, and S. Mathew. 2020. Influence of surface distresses on smartphone-based pavement roughness evaluation. *International Journal of Pavement Engineering*.

Similar methods are adopted to incorporate the acceleration data and recalibrate the model parameters, in order to remove the impacts due to the vibration of the vehicle body (for various vehicle types), operating speeds, etc.

The sampling frequency should be limited in the range of 80–120 hertz (Hz) when assessing smartphone roughness.[23]. The reason for that is when the sampling frequency is lower than 80 Hz vehicles need to travel at lower and constant speed, and at higher frequencies greater than 120 Hz, data storage and real-time processing can become an issue. This indicated that especially in urban areas where the speed limit is less than 50 km/h, the sampling interval should be kept at not greater than 300 millimeters (mm) along the traveled distance.[23]

This is further emphasized in Table 3, which shows the required speeds for different sampling frequencies to be compliant with the Class I, II, and III roughness measurement.

Table 3: Vehicle Operation Speeds for Different Smartphone Sampling Frequencies

Smartphone Sampling Frequencies (Hz)	Sampling Time Interval (s)	Speed Range to Satisfy a Class 1 Device, Sampling Interval ≤ 25 mm	Speed Range to Satisfy a Class 2 Device, 25 mm $\leq$ Sampling Interval ≤ 150 mm (km/h)	Speed Range to Satisfy a Class 3 Device, 150 mm $\leq$ Sampling Interval ≤ 300 mm (km/h)
40	0.025	≤ 3.6	≤ 21.6	≤ 43.2
80	0.0125	≤ 7.2	≤ 43.2	≤ 86.4
100	0.01	≤ 9	≤ 54	≤ 108
400	0.0025	≤ 36	≤ 16	≤ 432

Hz = hertz, km/h = kilometer per hour, mm = millimeter.

Source: Authors

It is also apparent that there are several external factors that would also influence the roughness values derived from a PSD-based approach; these would be incorporated during the model coefficient calibration to an extent. However, that would mean such models are not transferrable when the operating conditions under which calibration was done differently. Therefore, studies have been done to incorporate the other factors in the estimation of the IRI using smartphone-based acceleration data. Results of one such study that was done simulated vehicle movements on road profiles with different roughness values using the software Carsim, and the results of the multivariate regression analysis shown in Equation 4–6.[24] The data therein were used to develop a linear model for a large European van (LEV) (Eq. 4), a linear model for a D-class sedan (DSD) (Eq. 5), and a linear model which does not include the vehicle type (Eq. 6). However, since these were derived based on simulation results and subject to the define range of variables, the relative impact on the variables may differ from practical cases.

[23]   H. Jones and L. Forslof. 2014. Roadroid: Continuous road condition monitoring with smart phones. *Journal of Civil Engineering and Architecture*, 485–496.

[24]   G. Wang, M. Burrow, and G. Ghataora. 2020. Study of the Factors Affecting Road Roughness Measurement Using Smartphones. *Journal of Infrastructure Systems*.

$$IRI = 55.25 \times Grms - 0.07 \times Speed + 0.19 \times Npeop - 2.93 \times Stif - 1.09 \times DampF - 1.44 \times TyreS + 7.66 \quad (4)$$
$$R^2 = 0.83, SE = 1.39 \ m/km$$

$$IRI = 54.43 \times Grms - 0.06 \times Speed + 0.18 \times Npeop - 0.87 \times Stif - 0.89 \times DampF - 0.27 \times TyreS + 5.75 \quad (5)$$
$$R^2 = 0.79, SE = 1.54 \ m/km$$

$$IRI = 50.32 \times Grms - 0.06 \times Speed + 0.17 \times Npeop - 1.86 \times Stif - 0.9 \times DampF - 0.78 \times TyreS + 6.68 \quad (6)$$
$$R^2 = 0.74, SE = 1.69 \ m/km$$

Note:

Vehicle speed (Speed)	: 30, 50, 80
Vehicle type	: LEV, DSD
Sprung mass (number of people) (Npeop)	: 1, 2, 3 and 4 (70 kg per person)
Suspension stiffness (Stif)	: 0.8, 1, and 1.2
Suspension damping (DampF)	: 0.8, 1, and 1.2
Tire pressure (TireS)	: 0.7, 0.9 and 1
Road Profile (160 m length)	: Nine different road profiles

Machine Learning Approaches

Machine learning approaches offer the advantage of improving the accuracy of the model outputs compared to regression analysis-based approaches, as it is able to incorporate the complexities of how each independent variable is affecting the acceleration profile and its relationship with the roughness values.

The model types have been used successfully in related applications which predict road conditions from field measurements by several researchers as shown in Table 4.

Table 4 emphasizes that most machine learning models are able to adequately predict the roughness value that is comparable to conventional roughness measurement device outputs.

Table 4: Summary of Machine Learning Techniques Used for Roughness Evaluation

Case Study	Machine Learning Techniques Used	Description of the Roughness Evaluation	Results
Bajic et al.[a]	MLP k-nearest SVM ANN RF NB	On field test on a 50 km section. 23–38 features selected for the ML models based on the sequential feature selection and principal component analysis selecting subset of features than explain 99% of variance. Average IRI estimated for 100 m segments. In vehicle sensor acceleration readings obtained (frequency 50 Hz)	MLP best performing model, followed by SVM, RMSE 0.21. All models RMSE less than 0.24 IRI classification (low <0.9, medium 0.9–1.6, high >2.5) accuracy 74%
Wang, Burrow, and Ghataora[b]	MLP RF DRT	A detailed study on variables affecting the road roughness prediction using a smartphone inside of a moving vehicle in simulation settings. They studied the impact of different vehicle types and their mechanical properties and the driving styles on dynamical responses, as well as the position and the smartphone type. They developed multiple machine learning models	Best results of RMSE between 0.73 m/km and 0.91 m/km using the MLP model.
Adbelaziz et al.[c]	MLP	MLP model using the initial IRI, pavement age, rutting depth and cracks data	Prediction results obtained R^2 = 0.75 with actual IRI data
Jeong, Jo, and Ditzler[d]	CNN	IRI prediction, based on a half-car model simulation of vehicle dynamic response, at measured road profile and realistic driving speed profiles. Represented inputs as 2D images and utilized a 2D CNN model	Explored the CNN performance under different combinations of vehicles, inputs and driving speeds and concluded that they showed similar performances with the RMSE - 0.5–0.6 m/km

continued on next page

Table 4 continued

Case Study	Machine Learning Techniques Used	Description of the Roughness Evaluation	Results
Gong et al.[e]	RF	A random forest regression model for IRI prediction from the initial IRI (IRI at the moment of pavement construction), along with the rutting depth, road distress, traffic, climate, and pavement structure variables, was proposed.	Prediction results obtained $R^2 = 0.97$ with actual IRI data
Zhang et al.[f]	ANN	Applied for acceleration data of connected vehicles generated in a simulation. A sequential feature selection algorithm selected 76 features outputs for roughness estimation. Applied a convolution operation to extract common road-related features, while suppressing the impact of different mechanical properties and speeds of vehicles	IRI values as the target has an error of less than 0.3 m/km Qualitative classification accuracy 90%
Nitsche et al.[g]	MLP, SVM, RF	Weighted longitudinal profile was calculated (range and standard deviation) as the metric for roughness. A simulation framework with realistic road, driver, and vehicle models to calculate the response of a full-car vehicle model traversing different road models with specific roughness characteristics was used for the study.	ML models were effective in prediction roughness. SVM produced the best results.
Sollazzo, Fwa, and Bosurgi[h]	ANN	Levenberg-Marquardt algorithm was used to estimate the IRI using 25 hidden neurons, and the related records were randomly divided in the training (70%), validation (15%), and test (15%) groups.	The results shown the good precision with R^2 value fluctuates around 0.8.

continued on next page

Table 4 continued

Case Study	Machine Learning Techniques Used	Description of the Roughness Evaluation	Results
Zhou, Okte, and Al-Qadi[i]	RNN	RNN was implemented using PyTorch. The algorithm was constructed to predict the pavement's IRI using the time-series as learning data. Historical data of different AC pavement sections were extracted from the LTPP database. Independent variables for the analysis were ESAL, annual average temperature, annual total precipitation, annual total evaporation, annual rutting depth, annual fatigue cracking area, and annual transverse cracking length.	Using only one hidden layer with eight nodes was able to achieve the optimal model performance with the coefficients of determination as high as 0.77.

AC = asphalt concrete, ANN = artificial neural network, CNN = convolutional neural network, DRT = decision regression tree, ESAL= equivalent single axle load , IRI = international roughness index, Hz = hertz, k = nearest neighbor, km = kilometer, m = meter, ML = machine learning, MLP = multilayer perception, LTPP = long-term pavement performance, NB = naive bayes, RF = random forest, RMSE = root mean square error, RNN = recurrent neural network, SVM = support vector precision.

[a] M. Bajic et al. 2021. Road Roughness Estimation Using Machine Learning. *IEEE*.

[b] G. Wang, M. Burrow, and G. Ghataora. 2020. Study of the Factors Affecting Road Roughness Measurement Using Smartphones. *Journal of Infrastructure Systems*.

[c] N. Adbelaziz et al. 2020. International Roughness Index prediction model for flexible pavements. *International Journal of Pavement Engineering*, 88–99.

[d] J. Jeong, H. Jo, and G. Ditzler. 2020. Convolutional neural networks for pavement roughness assessment using calibration-free vehicle dynamics. *Computer-aided Civil and Infrastructure Engineering*, 1209–1229.

[e] H. Gong et al. 2018. Use of random forests regression for predicting iri of asphalt pavements. *Construction and Building Materials*, 890–897.

[f] Z. Zhang et al. 2018. Application of a Machine Learning Method to Evaluate Road Roughness from Connected Vehicles. *Journal of Transportation Engineering, Part B: Pavements*.

[g] P. Nitsche et al. 2012. Comparison of Machine Learning Methods for Evaluating Pavement Roughness Based on Vehicle Response. *Journal of Computing in Civil Engineering* .

[h] G. Sollazzo, T. Fwa, and G. Bosurgi. 2017. An ANN model to correlate roughness and structural performance in asphalt pavements. *Construction and Building Materials*, 684–693.

[i] Q. Zhou, E. Okte, and I. Al-Qadi. 2021. *Predicting Pavement Roughness Using Deep Learning Algorithms*. Transportation Research Record.
Source: Authors.

Several commercial mobile applications have been developed to estimate road roughness in an effective way. Most of these included the above-described roughness estimation techniques.

Summary

The following conclusions can be drawn with respect to roughness evaluation using smartphone-based data.

1. Converting vehicle body acceleration to PSD or root mean square (RMS) of vertical acceleration greatly facilitates the prediction of road roughness.
2. Vehicle body acceleration, smartphone type, vehicle type, and speed were found to be the dominant factors that influence road roughness measurement.

3.   A data collection system which utilizes smartphones suitably fixed inside a moving vehicle to assess road roughness can satisfy the frequency of data collection and accuracy requirements of a Class 3 device, if the vehicle speed is taken into account and a suitable data processing approach is adopted.

4.   Without knowing vehicle type and speed, resultant vertical acceleration or similar parameters to PSD can be utilized to assess road roughness to a similar degree of accuracy as can be achieved by a visual inspection (Class 4 roughness evaluation).

Smartphone-Based Data for Pavement Condition Evaluation: State of the Practice

Comparison of Existing Software for Smartphone-Based Pavement Condition Evaluation

Several smartphone applications have been developed to estimate pavement roughness.

Roadroid Mobile Application

This is a common application which was developed by Lars Forslof and Hans Jones in 2015 using accelerometer and GPS data collected from smartphones to estimate road roughness. This is the Google Android application that records vehicle vibration and correlates to the IRI. This application generates two outputs.

Estimated IRI (eIRI) - This output is based on a peak and RMS vibration analysis. The setup is fixed but made for three types of cars, and is thought to compensate for speeds between 20–100 km/h.

Calculated IRI (cIRI) - This output is based on the quarter-car simulation for sampling during a narrow speed range such as 60–80 km/h.

iRoad Application (UOM)

This used a machine learning approach to estimate road roughness by converting acceleration data into IRI values. In this approach, a mobile application is developed to capture the produced acceleration from a running vehicle by filtering noise and engine vibration.

All three acceleration components were considered here and found that vertical axis acceleration is more compatible with the variation of IRI which is simultaneously measured by using Bump Integrator Class III roughness measuring equipment. The GPS coordinates were used to locate the roughness variation on Google Maps, and both average vertical acceleration and RMS of x, y, and z acceleration were trained using an Euler angle-based algorithm to estimate the IRI as shown in Figure 35. Also, segmentation is done to check the estimation accuracy and found that a 100 meter (m) segmentation is more preferred rather than 200 m and 500 m segmentation. Moreover, it was found that IRI estimation using machine learning has significant accuracy with Mean Absolute Error of 0.65 compared with the Bump Integrator IRI.

Moreover, in machine learning approach, Random-forest algorithm is used in the backend to detect road anomalies (e.g., pothole or bump). The input was used as the vertical acceleration, horizontal acceleration which is perpendicular to the vehicle's moving direction, and vehicular speed as the features for the machine learning model. Also, a moving average filter for every feature to smooth them. A 0.93 precision of the anomaly detection was found from the machine learning approach. Also, a threshold-based approach was used to detect anomalies by checking the data points where vertical acceleration was greater than 10 m/s^2. However, only 0.85 precision was found by the threshold-based approach in anomaly detection.

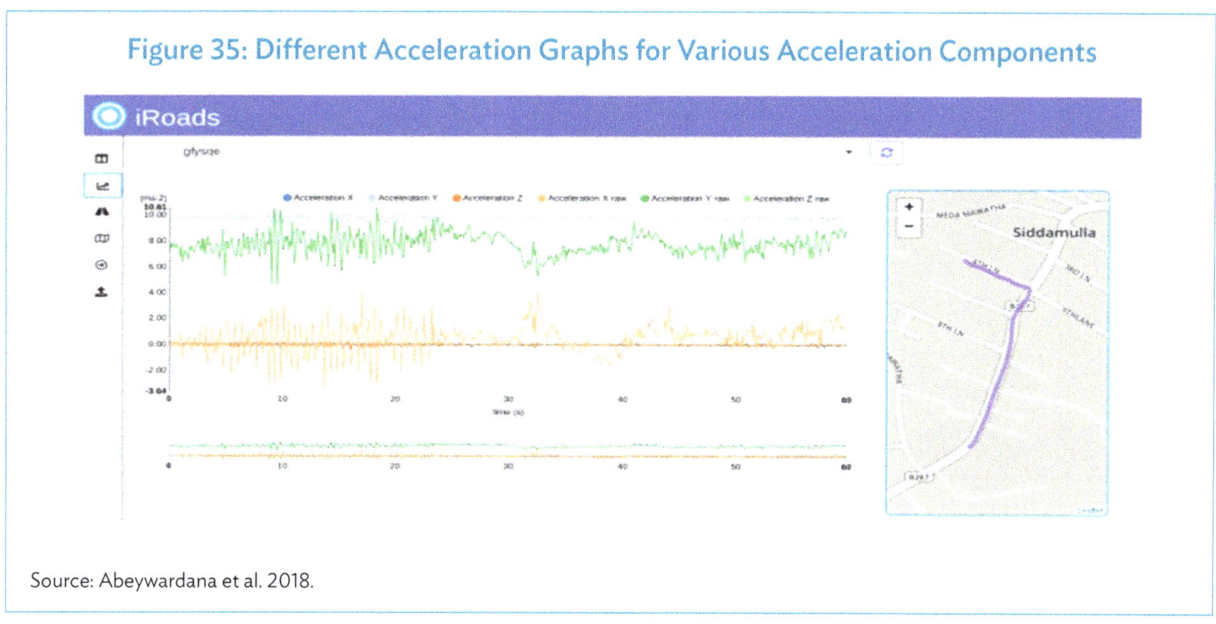

Figure 35: Different Acceleration Graphs for Various Acceleration Components

Source: Abeywardana et al. 2018.

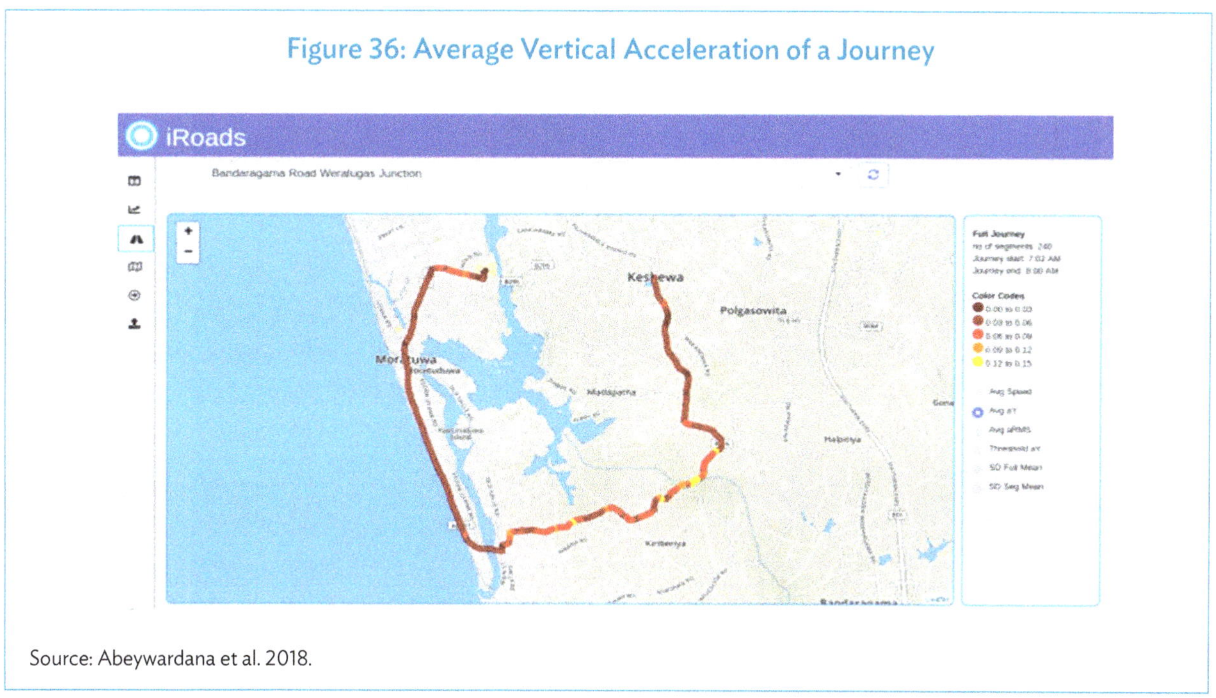

Figure 36: Average Vertical Acceleration of a Journey

Source: Abeywardana et al. 2018.

RoadLab Pro

The RoadLab application estimates road roughness based on kinematic and GPS sensors embedded in smartphones. In this method, vertical acceleration is directly correlated to road roughness. The roughness estimation is based on the vehicle suspension system and cellphone cradle basically while producing roughness value on time with a time lag for past 100 m section. In addition to the roughness estimation, RoadLab Pro can be used to map road networks, detect major road bumps, and also report road safety hazards. However, since it uses the correlation method, there would be some errors which are difficult to omit; for unpaved roads such as gravel and earth roads, this would not be a good option to measure road roughness.[25]

RoadBump

RoadBump uses accelerometer and GPS sensors to record vibrations and produce roughness either in terms of IRI or PSR with mapping output. RoadBump provides objective data on all types of roads such as asphalt, concrete, dirt, and gravel. Moreover, IRI graphs as a moving average or in segment lengths specified to 1 mile or km and presented as output. The main advantages of this application are ease of use, ability to customize the acceleration rate, and no need for network connection to record and review data.[26]

TotalPave

TotalPave is a smartphone-based system for collecting and reporting on objective road condition data. TotalPave consists of the following components:

- PCI Calculator: Mobile smartphone application used to collect and calculate Pavement Condition Index (PCI), a score between 0 and 100 measuring a road's surface distress.
- IRI Collector: Automated mobile application which streams raw smartphone sensor data to the cloud to be converted to IRI, a measure of a road's roughness between 0 and 10.
- Sidewalk Liability Manager: Mobile smartphone application used to report and document general sidewalk issues.
- Web Portal: Data management and visualization system accessed through the TotalPave website used to display and report on data collected using the three mobile applications.

---

25   SoftTeco. 2021. *SoftTeco Projects*. https://softteco.com/projects.
26   Grimmer Software. 2021. *RoadBump*. http://www.grimmersoftware.com/roadbump.html.

A summary of current smartphone technologies to measure roughness is given in Table 5.

Table 5: Summary of Current Smartphone Technologies to Measure Roughness

Product/Reference	Evaluation Metrics	Operational Features	Output
Roadroid	RMS of vertical acceleration	Multiple linear regression model to estimate IRI from RMS of vertical acceleration. The operational speed is 20–100 km/h with measuring frequency 100 Hz. Considered the factors affecting vehicle type, speed, tire pressure, sensitivity, etc.	IRI 70%-80% accuracy compared to the Class I type roughness measuring equipment.
RoadBump	Accelerometer and GPS sensors to record vibrations	Adoptive to all types of roads. Another main feature in this It can select the accelerometer cycle rates hence precision, and no need for network connection to record and review data	IRI
Byrne et al.[a]	Vehicle vertical accelerations	Acceleration processed using a band pass filter of 0.5–6 Hz. The effects of low vehicle speeds (5 km/h), cornering, and accelerating/decelerating are eliminated.	Pothole defects in terms of major and minor
StreetBump (MIT)	Three-axis acceleration	Threshold-based pothole detection approach is used. Developed as a web-based crowd sourcing to enable verifications by different drivers.	Pothole identification
RoadLab by World Bank	Vertical acceleration	Regression model to determine IRI from vertical accelerations. Considered the effect of the position of the smartphone, vehicle speed, and suspension type. On-time roughness estimation	IRI
iRoad Application (UOM)	Three-axis acceleration	Machine learning and threshold-based approaches to estimate IRI and vertical acceleration, and horizontal acceleration which is perpendicular to vehicle's moving direction used to detect pothole. Noise and engine vibration are filtered and IRI for 100 m section produced.	IRI Pothole detection
Bump Recorder[b]	Sprung acceleration data and position data from GPS	Bump Recorder estimates an unsprung elevation, and it assumes equal to road profile and calculates IRI. Measuring frequency of 100 Hz	IRI
Islam et al.[c]	Vertical acceleration	ProVAL program was used to estimate pavement roughness from the pavement profile. Uses a linear regression model to relate IRI to vertical acceleration	IRI
Douangphachanh and Oneyama[d]	Vertical acceleration	Data recording is done at an interval of 0.01 second or at a frequency rate of 100 Hz	IRI
Du et al.[e]	Z-axis accelerometer	A multilinear regression model was developed between IRI and the PSD of measured acceleration data	IRI (error is less than 15% in estimated and actual IRI)

continued on next page

Table 5 continued

Product/Reference	Evaluation Metrics	Operational Features	Output
Belzowski and Ekstrom[f]	Three axis accelerometer	Multi-regression models based on empirical data obtained from nine different smartphones at sampling frequency of 100 Hz.	IRI

GPS = global positioning system, Hz = hertz, IRI = international roughness index, km/h = kilometers per hour, PSD = power spectral density, RMS = root mean square.

[a] BM. Byrne et al. 2013. Identifying road defect information from smartphones. Road *and Transport Research*, 39–50.

[b] Y. Koichi. 2014. Collecting Pavement Big Data by using Smartphone. *1st IRF Asia Regional Congress Paper Submission Form.*

[c] S. Islam et al. 2014. Measurement of Pavement Roughness Using Android-Based Smartphone Application. *Transportation Research Record: Journal of the Transportation Research Board*, 30–38.

[d] V. Douangphachanh and H. Oneyama. 2013. A study on the use of smartphones under realistic settings to estimate road roughness condition. *Journal of the Eastern Asia Society for Transportation Studies.*

[e] Y. Du et al. 2014. Measurement of International Roughness Index by Using Z-Axis Accelerometers and GPS. *Mathematical Problems in Engineering.*

[f] B. Belzowski and A. Ekstrom. 2015. *Evaluating Roadway Surface Rating Technologies.* Michigan Department of Transportation.

Source: authors

- Roadroid https://www.roadroid.com/
- RoadBump http://www.grimmersoftware.com/roadbump.html
- Byrne et al. (2013) https://trid.trb.org/view/1255475
- StreetBump http://www.streetbump.org/
- RoadLab https://softteco.com/projects/roadlab
- iRoad Application (UOM) http://dilum.bandara.lk/wp-content/uploads/2019/06/iRoads-Project Report.pdf
- Bump Recorder https://www.bumprecorder.com/en/archives/1585
- Islam et al. (2014) https://doi.org/10.1061/9780784413586.053
- Douangphachanh and Oneyama (2013) https://doi.org/10.11175/easts.10.1551
- Du et al. (2014) https://doi.org/10.1155/2014/928980
- Belzowski and Ekstrom (2015) https://rosap.ntl.bts.gov/view/dot/29056

Review of Applications of Smartphone-Based Roughness Assessment

It is known that conventional roughness measurement equipment is producing pavement roughness accurately. However, as a cost-effective approach, smartphone-based roughness data also provide sufficiently accurate roughness data to assess pavement quality with significant repeatability.

Sandamal and Pasindu[27] has shown the applicability of smartphone-based roughness measurement into the rural low-volume roads by comparing smartphone roughness data with Bump Integrator IRI, which is a Class III type roughness measuring equipment. From that study, it was observed that it had a higher precision than the previous study with R-squared of 0.78 as shown in Figure 37.

---

[27] R. Sandamal and H. Pasindu. 2020. Applicability of Smartphone-based Roughness Data for Rural Road Pavement Condition Evaluation. *International Journal of Pavement Engineering.*

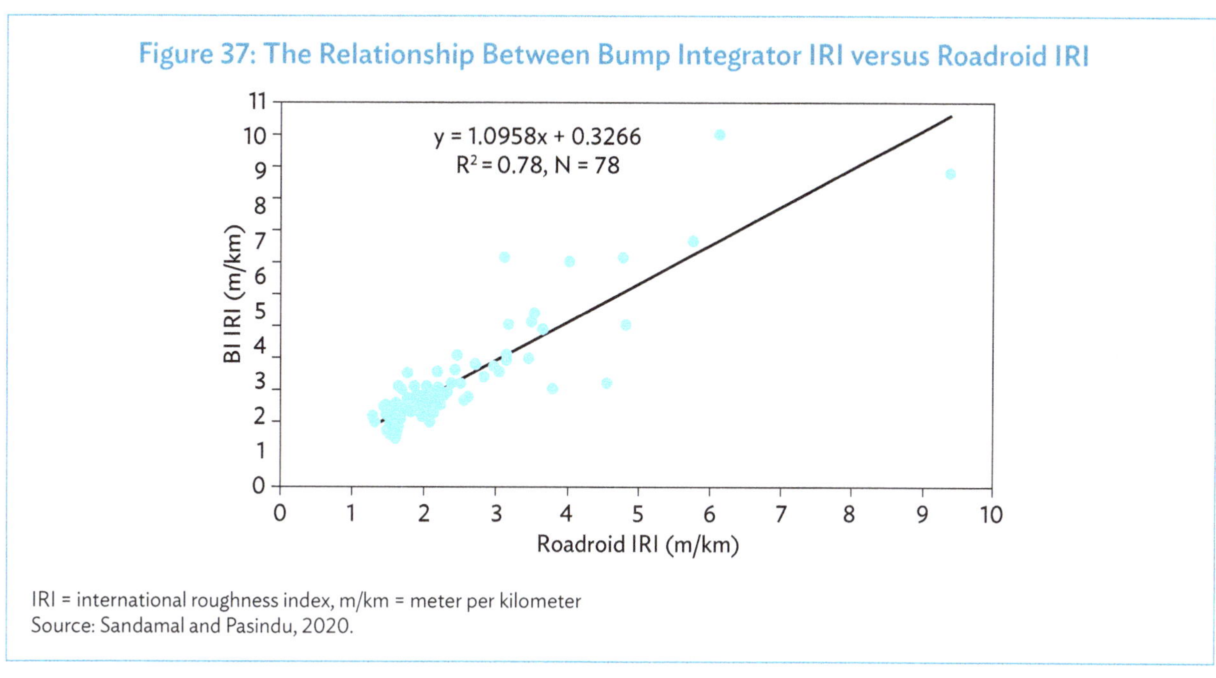

Figure 37: The Relationship Between Bump Integrator IRI versus Roadroid IRI

IRI = international roughness index, m/km = meter per kilometer
Source: Sandamal and Pasindu, 2020.

This roughness estimation was further continued to predict the pavement roughness based on the distress level. A relationship was developed by IRI (measured using smartphone-based app Roadroid) with the PCI, which is a parameter that deflects pavement quality based on the existing distress level. Figure 38 illustrates the derived relationship, while Equation 7 shows the IRI versus PCI function, which shows a good relationship with R² of 0.75. This model can be used even when there was no such system to measure road roughness especially for the local road agencies having resource constraints and lack of technical expertise.

$$IRI = \frac{10.98}{1+\left(\frac{PCI}{65.9}\right)^{4.27}} \tag{7}$$

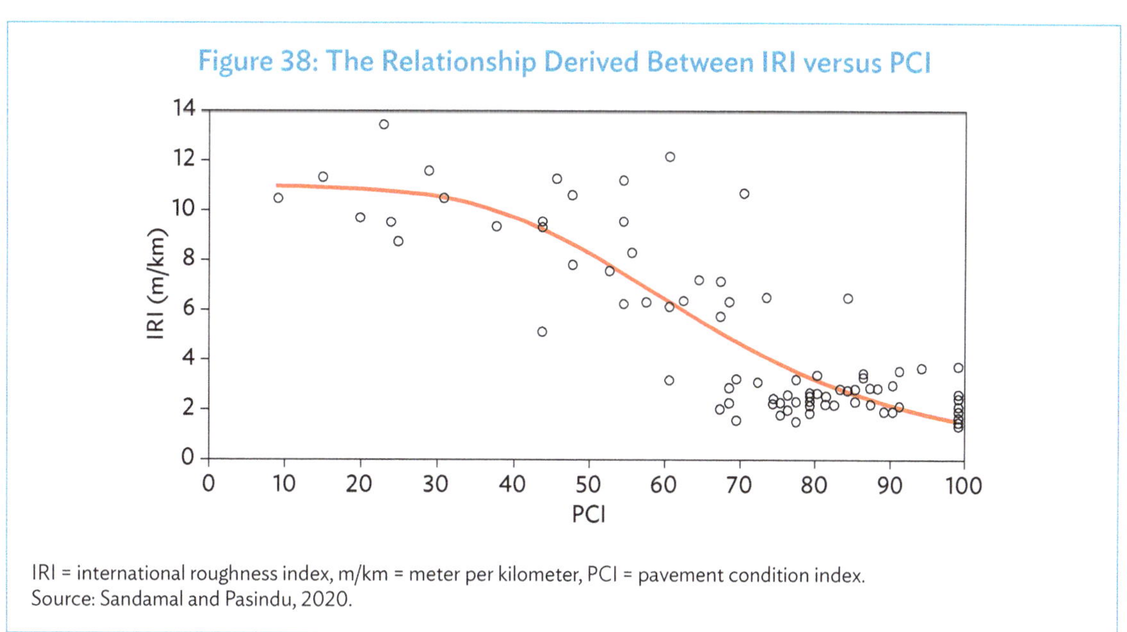

Figure 38: The Relationship Derived Between IRI versus PCI

IRI = international roughness index, m/km = meter per kilometer, PCI = pavement condition index.
Source: Sandamal and Pasindu, 2020.

Thube, Jain, and Parida[28] have shown that correlation between IRI and resultant acceleration ranges from 0.55 to 0.75 for different vehicle speeds. There is also no significant variation with change in speed of a vehicle. Moreover, Gonzales at al.[29] used tri-axial accelerometer on a car to obtain acceleration information and estimate road surface roughness.

TotalPave[30] conducted a study to validate the smartphone-based roughness with precise IRI measurements using TotalPave IRI versus laser profiler IRI (Hawkeye 2000). It was found that TotalPave is within about 10% of a Class 1 laser profiler. This means that all things being equal, controlling for lateral variability and other random sources of error, in a one-to-one comparison a single IRI value collected with TotalPave was within 10% of the Class 1 profiler.

Yeganeh et al.[31] evaluated the applicability of smartphone-based roughness measurement with pavement condition indices such as ride quality index (RQI) and pavement distress index (PDI). It was found that smartphone IRI significantly reflects the pavement condition while indicating good relationship with PDI and RQI having R^2 values of 0.5 and 0.81, respectively.

$$IRI = -970 \text{ PDI} + 1317.97 \qquad (R^2=0.52) \quad (8)$$

$$IRI = -1.27 \text{ RQI} + 8.95 \qquad (R^2=0.81) \quad (9)$$

Expanding Smartphone-Based Data to Crowdsourced Data for Pavement Condition Assessment

Introduction to Crowdsourced Data Analytics

Crowdsourcing data refers to data, information, or opinions collected from a large group of users who submit their data by using either the internet, social media, or smartphone apps.[32] Crowdsourcing is an emerging concept in the data mining field, while a sourcing model is designed to obtain contributions from individuals from a large community over the internet to gather new data revolving around that model in the form of big data. However, crowdsourcing data may involve ambiguous and contradictory data from the participants. Therefore, it is essential to interpret the real meaning of data by developing data screening tools. Crowdsource data analytics comprise of several processes as shown in Figure 39.[33] Thus, the rich knowledge and information provided by researchers will facilitate highly efficient data.

28   D. Thube, S. Jain, and M. Parida. 2007. Development of PCI Based Composite Pavement Deterioration Curves for Low Volume Roads in India. *Highway Research Bulletin, Indian Roads Congress.*

29   A Gonzales et al. 2008. The use of vehicle acceleration measurements to estimate road roughness. *Vehicle System Dynamics,* 483–499.

30   TotalPave. 2021. *Laser Profiler vs Smartphone – Precise vs Accurate IRI Data.* Retrieved from TotalPave: https://totalpave.com/blog/laser-profiler-vs-smartphone-precise-vs-accurate-iri-data/

31   S. Yeganeh et al. 2019. Validation of Smartphone-Based Pavement Roughness Measures. Cornell University.

32   Dotdash Meredith publishing family. 2021. *Investopedia .* Retrieved from Crowdsourcing: https://www.investopedia.com/terms/c/crowdsourcing.asp

33   X. Yang et al. 2017. Copula-based Multi-dimensional Crowdsourced Data Synthesis and Release with Local Privacy. *GLOBECOM 2017 - 2017 IEEE Global Communications Conference.* IEEE.

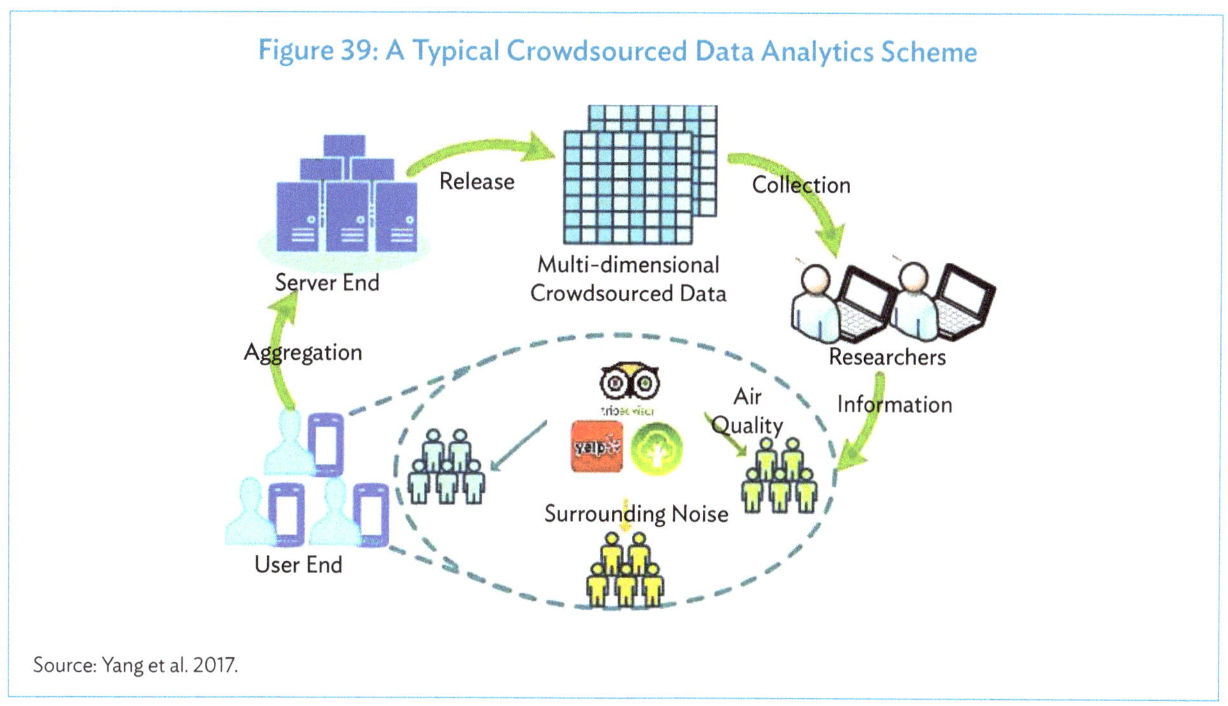

Source: Yang et al. 2017.

Crowdsourced Data for Pavement Condition Assessment

Various indicators have been adopted to measure pavement condition. However, there are only a few parameters that can be used to collect data by using crowdsourced data sources. Sensor data obtained from the moving vehicle or connected vehicle, image, or video-based data are known as common crowdsourced data types.[34] Once the data is collected using the platform provided, the next steps involve data storage and analysis. The stored data is used to perform the analysis using machine learning techniques such as CNN (footnote 34).

Yang et al (footnote 33) illustrated the different crowdsourced data types required to evaluate the various pavement condition assessment parameters. Different performance metrics are identified with those data; as an example, a pothole identification is based on the acceleration data, and this creates and logs geocoded "event" at a point where the data indicate that the vehicle has experienced sudden acceleration due to a bump or pothole. A number of vehicles collecting such data could be used to highlight rough pavement and potholes.

Moreover, primarily to measure roughness, accelerometer and vehicle system operations data can be used to gauge the interaction of the vehicle with the pavement. Obtaining a standard roughness metric, such as IRI, generally requires advanced measurement equipment. However, the unevenness can be calculated/convertible using the number of probe vehicles data-driven in a specific road stretch. Denis showed that even with a best-case scenario, such a measurement would still not be as precise as a standard like IRI, but has been shown to correlate to IRI significantly.

---

[34]   E. Denis. 2014. Pavement Condition Monitoring with Crowdsourced Connected Vehicle Data. *Transportation Research Record Journal of the Transportation Research Board*, 1 - 36.

A case study conducted by Jeong, Jo, and Ditzler[35] in Arizona showed that the incorporation of deep learning (i.e., CNN) with pavement vehicle interaction principles and multimetric smartphone sensor technologies enabled accurate pavement condition assessment from daily driving of anonymous (i.e., no need for calibration) vehicles. Figure 40 shows the developed framework to convert crowdsourced data into IRI value.

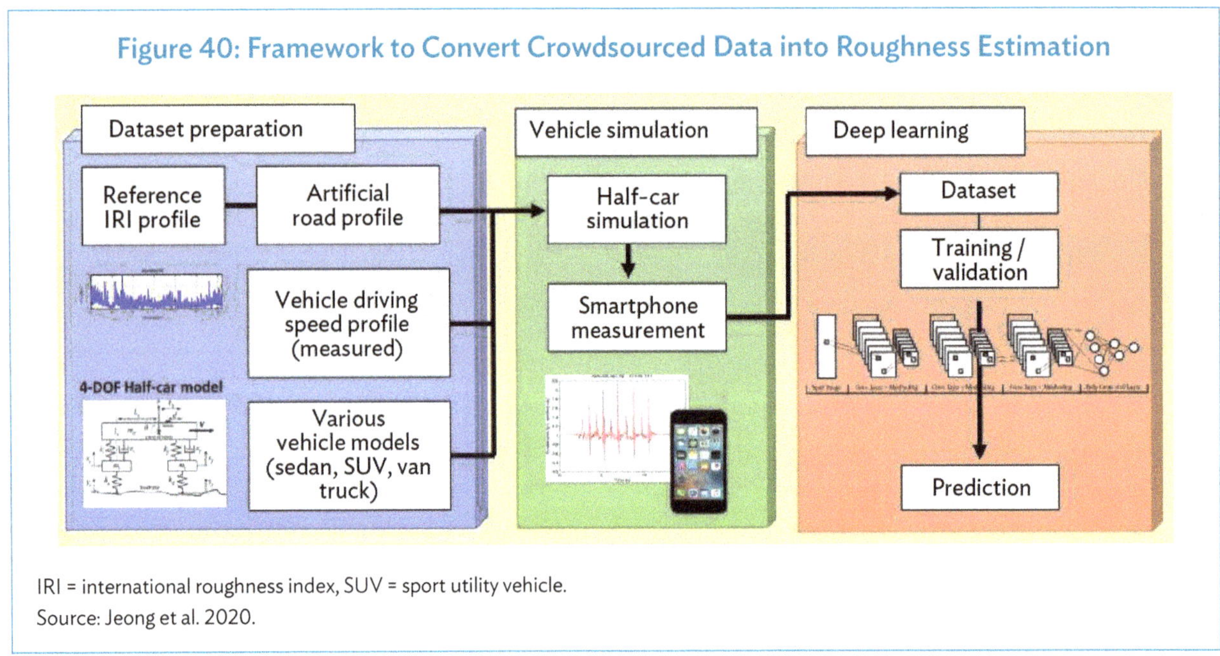

Figure 40: Framework to Convert Crowdsourced Data into Roughness Estimation

IRI = international roughness index, SUV = sport utility vehicle.
Source: Jeong et al. 2020.

The smartphone measurement data can be converted into two-dimensional or 2D image and the CNN application detect key feature from data use to filters by utilizing driving speed, vertical acceleration and angular velocity (pitch motion), etc.

However, the most common method of adopting crowdsourced data into pavement condition assessment is related to the roughness estimation. In this practice, it is essential to filter surface defects such as bumps and potholes on the road surface through built-in GPS and accelerometer sensors, and interpret the general roughness from unevenness. There are several solutions proposed by research studies such as (i) retrieve GPS and accelerometer data from the smartphone, and (ii) estimate the RMS values of the acceleration data perpendicular to the screen plane, and (iii) detect the number of transient events and evaluate the overall quality of the road surface roughness.[36]

The vertical acceleration component provides roughness measurement, by filtering the transient events these crowdsourced-based acceleration reading can be converted into a roughness value easily. Figure 41 shows illustrative example of how accurately it would be able to predict the IRI value and road defects by using the data collected from crowdsourced data.

---

35   J. Jeong, H. Jo, and G. Ditzler. 2020. Convolutional neural networks for pavement roughness assessment using calibration-free vehicle dynamics. *Computer-aided Civil and Infrastructure Engineering*, 1209–1229.

36   X. Li, R. Chen, and T. Chu. 2014. *A Crowdsourcing Solution for Road Surface Roughness Detection Using Smartphones*. Texas A&M University.

Figure 41: An Illustrative Example of Predicting IRI and Pothole Using Crowdsourced Acceleration Data

	Road Segment A	Road Segment B	Road Segment C
Mean RMS m/s^2	0.2233	0.3183	0.7824
Transient /km	0	3	15
Raw RMS			
Smoothed RMS and Transients			
Road Image			

IRI = international roughness index, km= kilometer, m/s^2 = meter per square meter, RMS = root mean square.
Source: Li et al. 2014.

The parallel development of higher-fidelity sensors, light detection and ranging (LiDAR), and efficient machine learning algorithms has enabled collecting pavement condition data at a high resolution and in real time. Nowadays, most of the high-end connected and semi automated vehicles are equipped with powerful sensors and high-definition cameras, LiDAR. Therefore, that data can be used to update agency asset management systems with the latest information.

For example, a vehicle can use machine learning to process an incoming image from an on-board camera to recognize road defects such as pavement cracks, potholes, pavement markings, signs, etc. As vehicles collect data on these assets, they can compare them over time to determine if they are deteriorating at an expected, slower, or faster rate. However, these data attributes vary based on technology and specific asset focus, but generally, data includes location, time stamp, and asset information. Asset information can be available as point clouds, images, deviation measurements, etc. In the case of collecting the data, the following attributes have to be considered:

- Latency: Certain types of data collection methods can be close to real time. For example, camera identification of specific features using image processing and machine learning is quick and is available shortly after.
- Details: High-resolution data generally have a high level of detail. These methods involve the collection of high-resolution images or billions of point measurements transformed into very high-quality asset information.

- Quality and coverage: Quality of high-resolution data is associated with high cost, which affects the level of coverage. For example, LiDAR scans can be expensive and limit the extent of coverage.

Figure 42 shows the sample of pavement data collection that can be achieved with the high-resolution camera data as an emerging crowdsourced data type.[52]

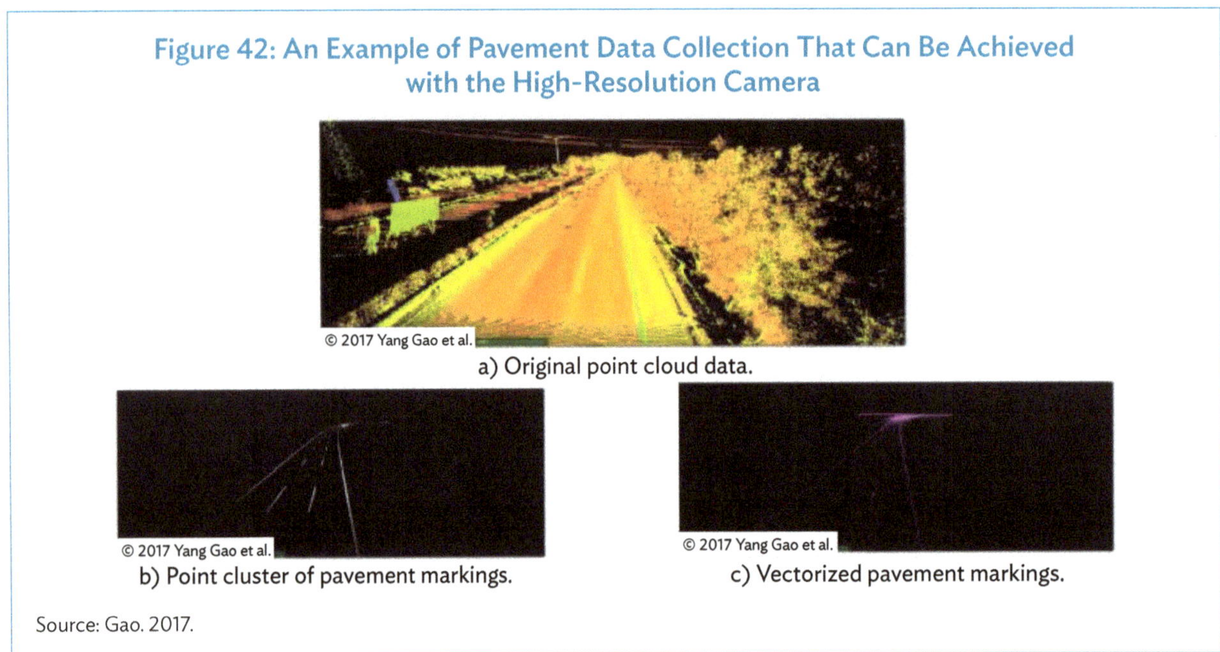

Figure 42: An Example of Pavement Data Collection That Can Be Achieved with the High-Resolution Camera

a) Original point cloud data.

b) Point cluster of pavement markings.

c) Vectorized pavement markings.

Source: Gao. 2017.

Smartphone-Based Platforms for Pavement Defects Reporting

With the emergence of smartphone-based technologies into the road asset planning, pavement defects such as potholes and cracks can be reported using smartphone-based platforms. These defects are captured either using accelerometer data, 2D or 3D image/video data, etc. For example, the detection of the potholes using smartphones were carried out in the Street Bump project[38] in Boston, Massachusetts in the US. Over time, various researchers have developed similar methods to monitor the quality of the road surface and the subsequent detection of critical points through the combined use of most sensors inside smartphones. In this section, some of the available platforms or applications are described which have the capability to detect pavement defects.

---

37   Y. Gao. 2017. Automatic extraction of pavement markings on streets from point cloud data of mobile LiDAR. *Measurement Science and Technology*.

38   Street Bump. 2020. Where's Street Bump being used? http://www.streetbump.org/about

Roadroid

The Roadroid[39] app offers simple and innovative data collection procedures in addition to roughness analysis. The Roadroid Image Analysis module can be used for automated classification of various properties/road parameters based on survey photos uploaded to the Roadroid app. Currently, the system supports classification of road dust levels using two different algorithms, corrugation, pothole and loose gravel on a gravel road. Figure 43 represents such locations on a map which helps to identify the distressed locations.

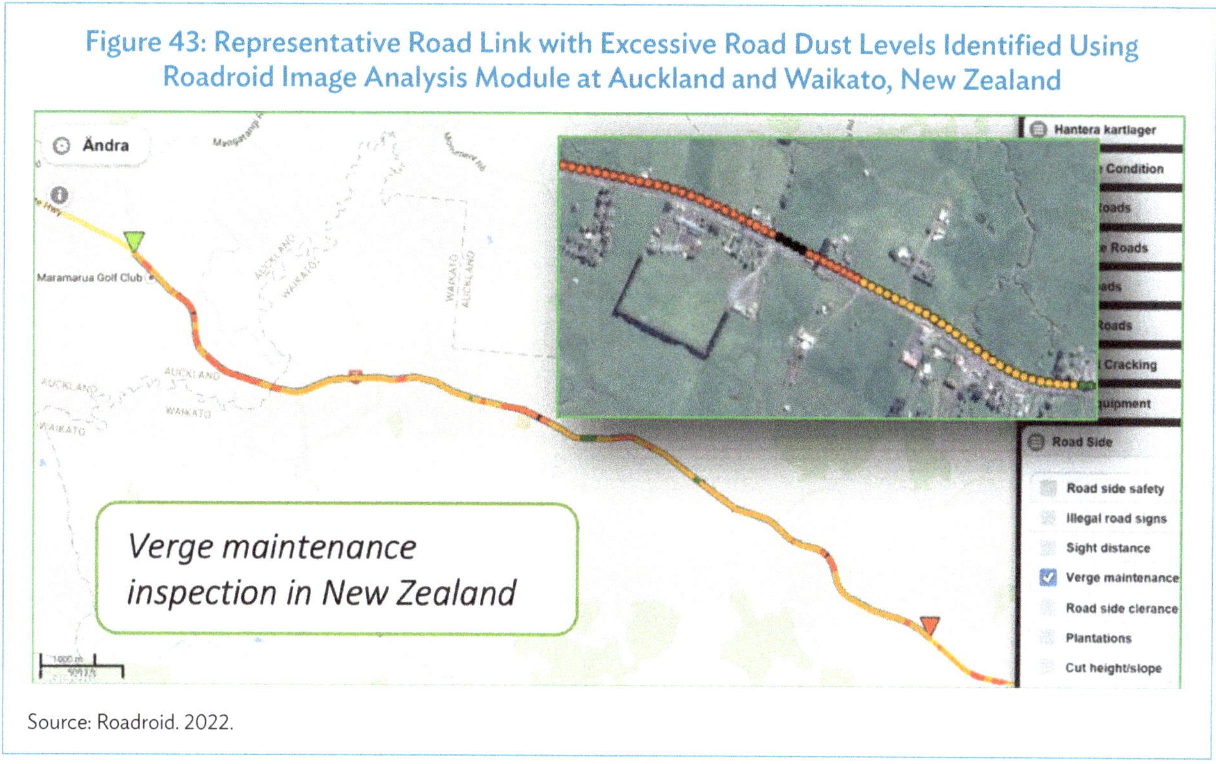

Figure 43: Representative Road Link with Excessive Road Dust Levels Identified Using Roadroid Image Analysis Module at Auckland and Waikato, New Zealand

Source: Roadroid. 2022.

Street Bump

Street Bump[40] is a project in Boston, where it allows road condition data collection while driving. It aggregates the data across users to provide the city with real-time information to fix short-term problems and plan long-term investments. Street Bump utilizes two of the phone's sensors, its accelerometer and GPS. The sensors detect "bumps" that it later maps. A number of factors determine when an individual bump is recorded such as the type of car and the speed of the vehicle. After uploading trip sensor data to the app's server, the number of potholes and their location can be identified.

---

39   Roadroid. 2022. Roadroid. https://www.roadroid.com/
40   Street Bump. 2020. Where's Street Bump being used? http://www.streetbump.org/about

RoadData

The RoadData[41] mobile app uses GoPro cameras to obtain high-definition video of a road and has the ability to calculate a 0–100 rating for each road. It features an automated distress identification module that utilize machine learning and computer vision algorithms to look at each road section and identify specific types of cracking and potholes (Figure 44). Moreover, it includes Street-smart, a cloud-based platform that provides an overview of the distress density across a road network (Figure 45).

Figure 44: The Automated Cracking and Pothole Identification Module (RoadData)

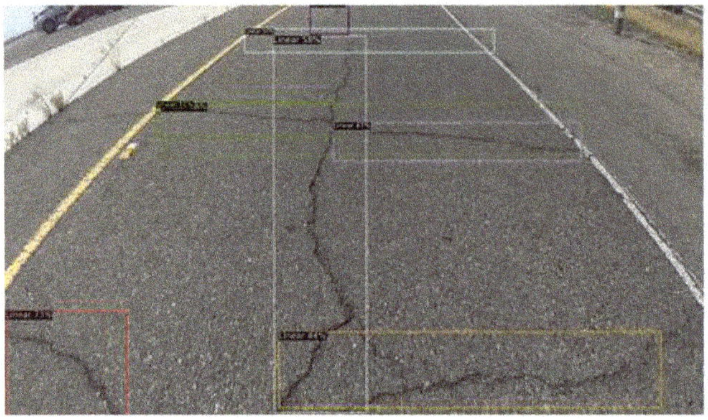

Source: RoadData. 2022.

Figure 45: Street-Smart Cloud-Based Platform to Report Distresses, Webster, Texas

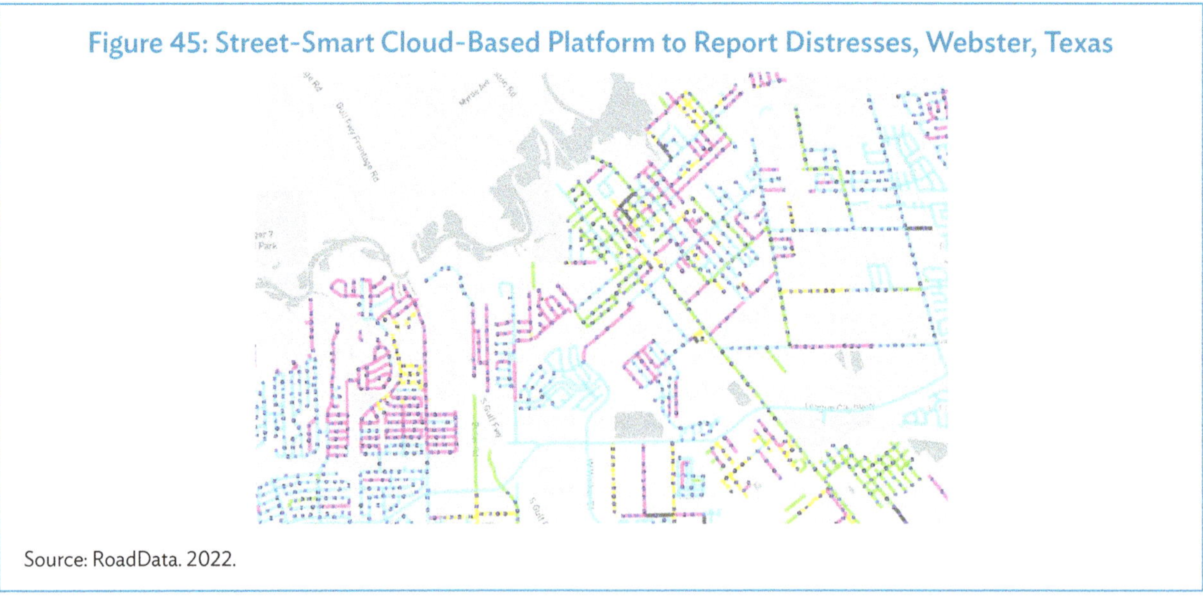

Source: RoadData. 2022.

---

[41]   RoadData. 2022. *RoadData - measuring highway and motorway markings for retro-reflectivity information you can trust.* https://roaddata.co.nz/?page=home.

TotalPave

In the TotalPave[42] app there is a separate module called "TotalPave PCI" which enables users to compute on-the-spot PCI values for any section of road. Only pavement distress type and dimensions are required as inputs which are ASTMD6433 compliant. Information collected using the application is streamed wirelessly to space reserved for the customer on TotalPave's cloud-based server.

Figure 46: TotalPave Portal Home Screen, Montana, United States

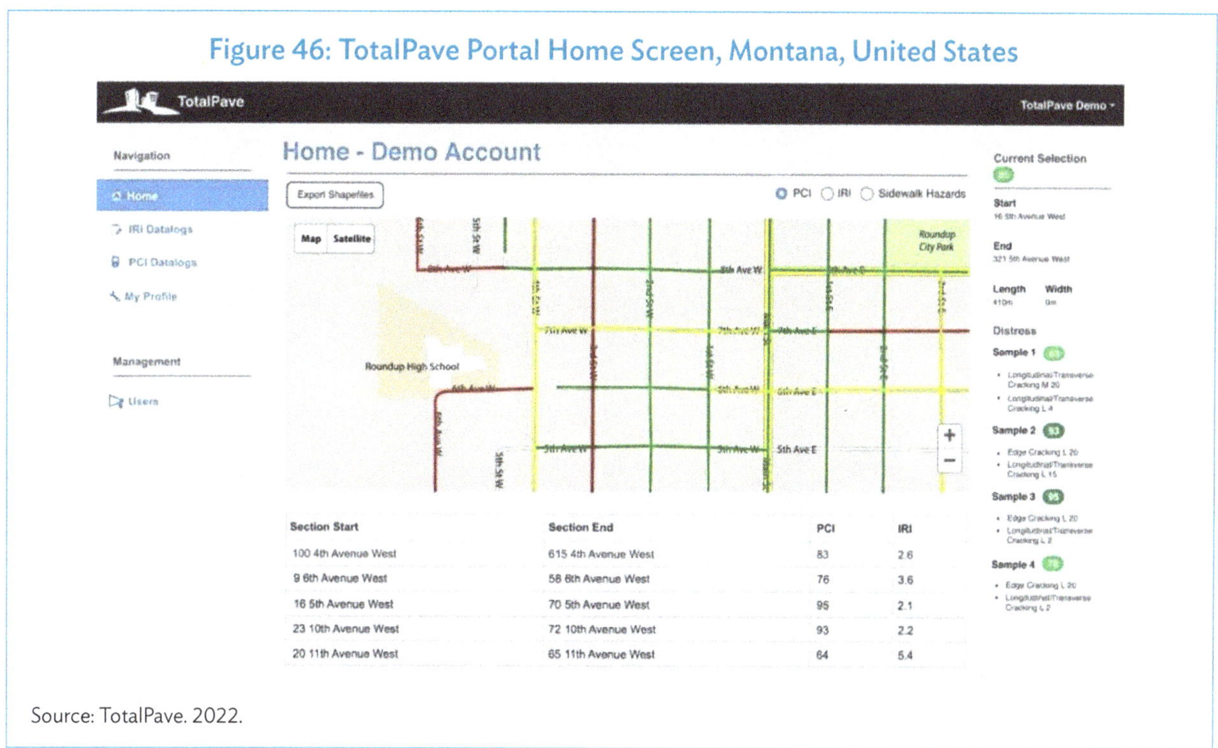

Section Start	Section End	PCI	IRI
100 4th Avenue West	615 4th Avenue West	83	2.6
9 6th Avenue West	58 6th Avenue West	76	3.6
16 5th Avenue West	70 5th Avenue West	95	2.1
23 10th Avenue West	72 10th Avenue West	93	2.2
20 11th Avenue West	65 11th Avenue West	64	5.4

Source: TotalPave. 2022.

RoadBotics

RoadBotics[43] empowers governments and communities to make objective, data-driven decisions about their road networks by using automated image mapping and road assessments, generating interactive maps, unbiased ratings, and practical tools to save time and cost. RoadBotics started with an advanced AI product capable of automatically assessing roadways.

The automated pavement assessment process for RoadWay has blueprinting, data collection, AI analysis, and delivery of the results on RoadWay. Using ASTM D6433 as the standard for defining and assessing pavement distresses, RoadBotics's AI model was trained to identify 18 distresses that fall into six categories: Potholes, Fatigue Cracking, Pavement Distortion, Patches, Transverse/Longitudinal Cracking, and Surface Deterioration. In the analysis, for each 10-foot section of the road being analyzed, the presence and/or pervasiveness of one or multiple pavement distresses will factor into the pavement rating. Pervasive distressing will result in a poor rating, whereas newly paved roads will have a good rating.

---

42   TotalPave. 2022. *Collect IRI and PCI with Your Smartphone.* https://totalpave.com/
43   RoadBotics. 2021. *Visualize and analyze your roads and other infrastructure assets .* https://www.roadbotics.com/

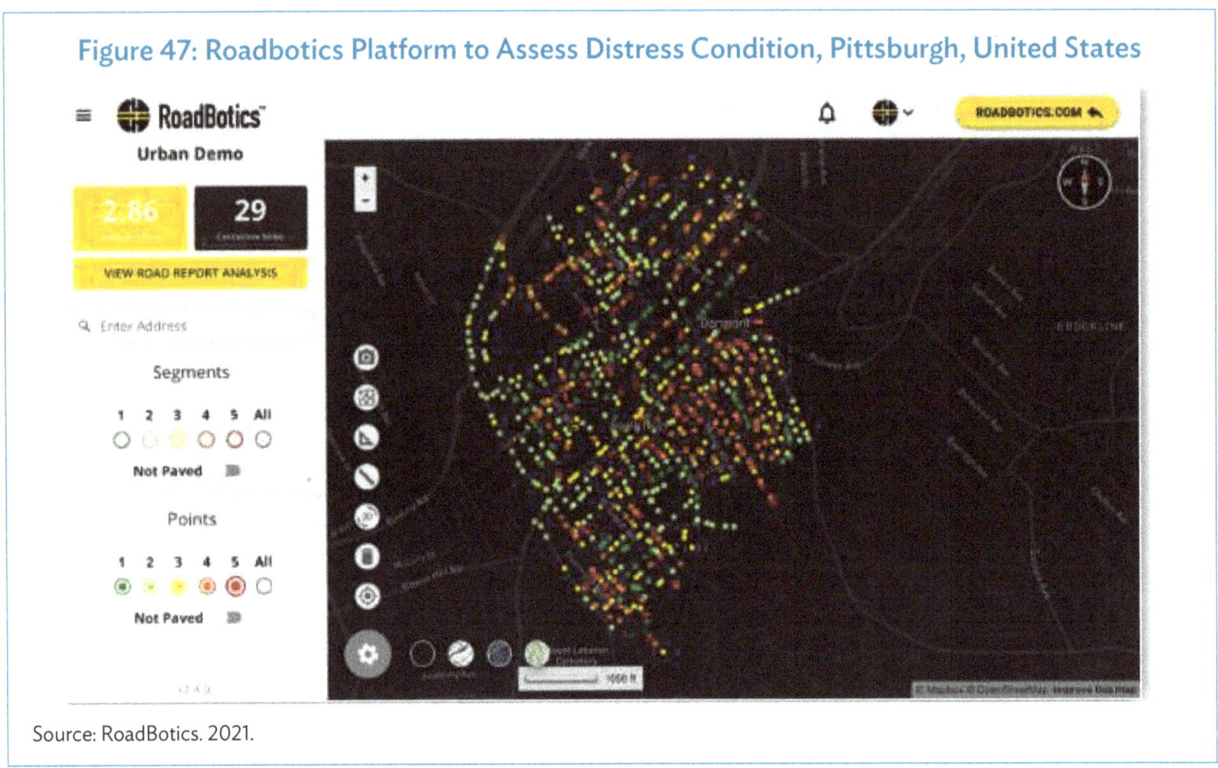

Source: RoadBotics. 2021.

Application of Google Traffic Data for Road Condition Evaluation

Pavement surface condition is one of the factors that affect driver comfort, operation speed, service volume, and traffic safety. It is understood that there is an impact from road roughness on speed patterns for different roadway segments under different traffic flow levels.

Travel time data can be obtained from Google Distance Matrix API, which is a processed information released based on crowdsourced mobile phone data. A cloud-based data acquisition platform has the ability to collect data by accessing the Google Distance Matrix API. The travel time observed from Google Distance Matrix API is then verified with the travel time information collected by using GPS-enabled probe vehicles. The results from Kumarage[44] indicated that there is a significant agreement between the travel time given by Google Distance Matrix API and actually observed data for both short-distance and long-distance trips. Finally, this data can be converted into Google speed data that can be used to establish the different speed components such as free flow speed (FFS), average speed (V_{50}), etc.

This FFS, V_{50} has a strong relationship between road condition especially with consistent pavement condition evaluation parameter such as IRI.[45] However, there can be some variation of the measurement speed (from Google traffic data) with IRI (road condition measurement metrics) when the probe vehicle differs. Eq. 10 and 11 show examples for such difference which is illustrated from a case study done in India (Chandra, 2004).

[44] S. Kumarage. 2018. *Use Of Crowdsourced Travel Time Data In Traffic Engineering Applications.* Moratuwa: University of Moratuwa.

[45] C. Abeygunawardhana, R. Sandamal, and H. Pasindu. 2020. Identification of the Impact on Road Roughness on Speed Patterns for Different Roadway Segments. *Moratuwa Research Conference (MERCon)*, 425 - 430. Moratuwa: IEEE.

$$V_{ssfc} = 66.9 - 0.0034*UI \qquad (R^2=0.51) \quad (10)$$

$$V_{ssfhv} = 51.6 - 0.0019*UI \qquad (R^2=0.50) \quad (11)$$

Where, V_{ssfc} – Speed of car in km/h , V_{ssfhv} – Speed of heavy vehicle in km/h , UI – Unevenness index in mm/km

However, this relationship shows that the difference is comparatively low when compared with the relationship with speed versus road condition. Moreover, road roughness is shown to have a significant relationship with FFS and V_{50} when the increase in IRI is beyond the value of 7 m/km[60] as shown in Figure 48 and 49.

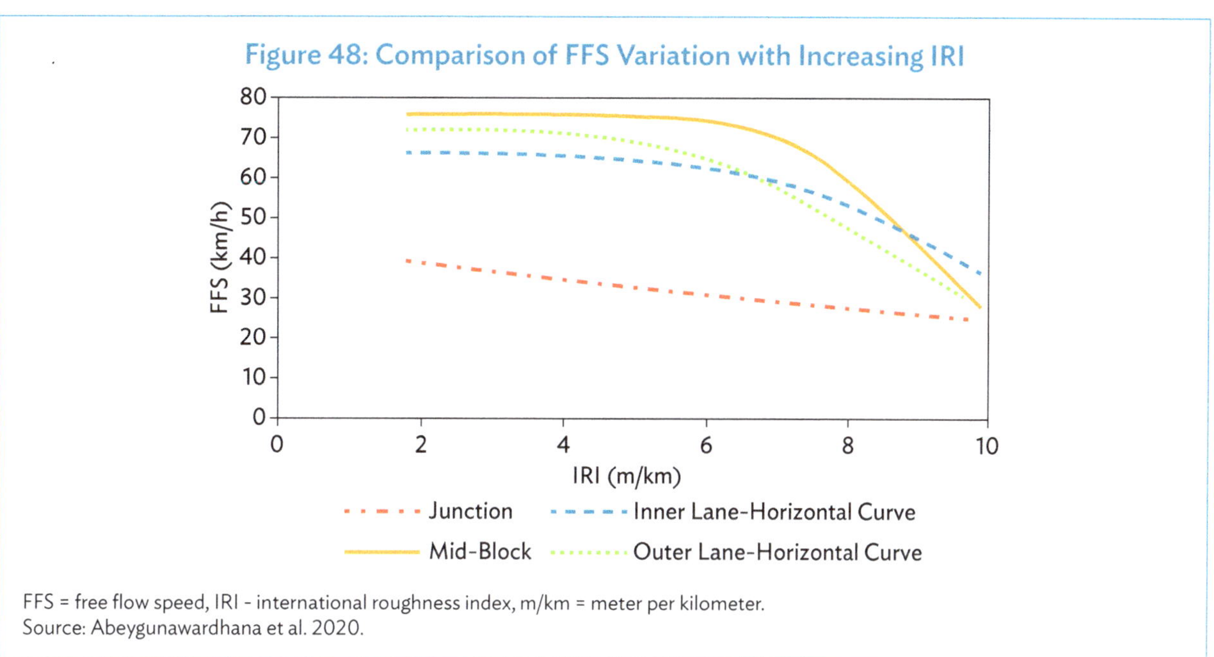

Figure 48: Comparison of FFS Variation with Increasing IRI

FFS = free flow speed, IRI - international roughness index, m/km = meter per kilometer.
Source: Abeygunawardhana et al. 2020.

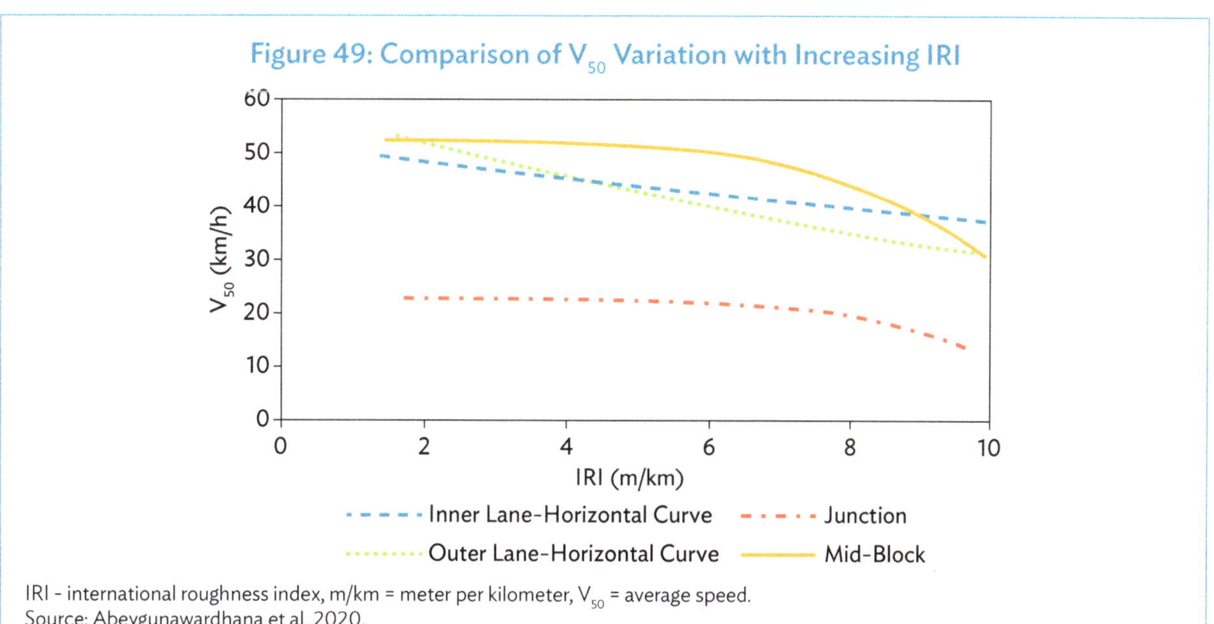

Figure 49: Comparison of V_{50} Variation with Increasing IRI

IRI - international roughness index, m/km = meter per kilometer, V_{50} = average speed.
Source: Abeygunawardhana et al. 2020.

This suggests that the Google traffic data is helpful for road agencies in their decision-making process on maintenance. The Google data available through crowdsourced data sources, using appropriate filtering on such data, can be used to predict the condition.

Requirements in Adopting Smartphone-Based Pavement Condition Assessment

The use of smartphone technology as a tool to collect data is a viable alternative due to its low cost and easy-to-use features, in addition to its potentially wide population coverage as probe devices. Modern smartphones are equipped with many useful sensors such as accelerometers, gyroscope, and GPS. Many researchers have proposed approaches to use smartphones to measure pavement roughness and, detect road anomalies in the hope of increasing the ability of road authorities to collect data at an appropriate frequency required for management and maintenance planning. However, there are many requirements in adopting smartphone-based pavement condition assessment into the pavement asset planning by replacing conventional methods.

Modern smartphones consist of various embedded sensors along with on-board storage, computing, and communication capabilities. Owing to these features, they have become an intelligent, scalable, autonomous, and potentially cost-free component of the next-generation civil infrastructure monitoring system. An emphasis is placed on the sensing capabilities of smartphones and their crowdsourcing power for monitoring road networks.

As the primary factor, there should be access to smartphones with features such as accelerometers, microphones, GSM radio, and GPS to monitor road and traffic conditions, and also significant strength of the mobile signal to capture the roadway data accurately. The data collection rate is another important aspect to consider, even though it depends on the user. In most of the researchers, the data collection is conducted at 10–100 samples per second.[46] However, the higher the data collection rate, better the accuracy of the estimation (roughness or pavement defects).

Moreover, there is a major concern when we adopt smartphone-based applications for the roughness measurement such as the speed of the survey. Generally, when the speed is less than 20 km/h,[47] it is difficult to get a complete dataset, hence operation speed usually tends to keep between 20–100 km/h. In addition, a large number of factors need to be taken into account when determining road conditions from smartphone-based systems. These factors are in categories of vehicle-related aspects and the form of the road surface condition.

However, in some smartphone models, issues with the sensors may causes random errors due to manufacturing problems such as the misalignment of the sensor (causing a bias error in measuring acceleration) and temperature-related sensitivity.[48] Moreover, the number of processes running on a smartphone can increase its internal temperature and in turn lead to errors ranging from 1 to 2.4 mg/°C when converting the physical accelerating force to an electric charge via the smartphone's microelectromechanical quantified the sensitivity to gravity of some common smartphone types. They found that for older smartphones (e.g., iPhone 4 and LGNexus 5), the error in measuring gravity

---

[46]   Bisconsini, D. R., Nunez, J. Y., Nicoletti, R., & Junior, J. L. 2018. Pavement Roughness Evaluation with Smartphone. *International Journal of Science and Engineering Investigations, 7*, 43–52.

[47]   Sattar, S., Li, S., & Champmon, M. 2018. Road Surface Monitoring Using Smartphone Sensors: A Review. *Sensors.*

[48]   Woodman, O. J. 2007. *An introduction to inertial navigation.* https://www.cl.cam.ac.uk/techreports/UCAM-CL-TR-696.pdf

was between 0.0164 and 0.025 g, and higher in more modern smartphones,[49] such as the iPhone 6 system.[50]

Driver behavior has a significant effect on smartphone-based data collection especially when using sensor data. Therefore, throughout the survey period, constant driver behavior condition must be maintained to get consistent measurement. The influence of driving style on simulated vehicle body acceleration was investigated by simulating a D-class sedan (DSD) traveling on a road with IRI 5 m/km (i.e., an older road in fair condition) for the five different regimes (Table 6). Compared with driving at a constant speed, accelerating or decelerating affected the simulated body acceleration between 3% and 7%.

Table 6: Effects of Driving Regimes on Simulated Grms

Acceleration/Braking	Speed Ranges (km/h)	Traveled distance (m)	Time (s)	Grms (g)	Difference Compared with Driving at Constant Speed (%)
Constant acceleration (20% throttle control)	0–78	160	11.7	0.083	–2.71
Constant driving speed	49	160	11.7	0.085	–
Half-throttle acceleration	0–100	160	10.0	0.087	–6.88
Constant driving speed	58	160	10.0	0.092	–
Deceleration at constant brake pressure	78–0	152	15.9	0.070	4.95
Constant driving speed	35	152	15.9	0.067	–
Constant speed 100 km/h and braking after 2 seconds	100–0	137	7.7	0.091	–5.96
Constant driving speed	64	137	7.7	0.096	–
Full throttle acceleration, then braking	0–82, then to 62	160	10.7	0.085	–5.55
Constant driving speed	54	160	10.7	0.090	–

Grms = root mean square acceleration, km/h = kilometer per hour.
Source: Based on author's analysis.

Therefore, it is essential to consider on the driving style when adopting smartphone-based pavement assessment.

---

[49] Kionix. 2015. *AN012 accelerometer errors.* http://kionixfs.kionix.com/en/document/AN012%20Accelerometer%20Errors.pdf.

[50] A. Kos, S. Tomažič, and A. Umek. 2016. Evaluation of smartphone inertial sensor performance for cross-platform mobile applications. *Sensors*, 477–492.

Feasibility of Adopting Smartphone-Based Assessment Methods

Applicability for Road Condition Assessment and Maintenance Decision-Making

Widespread acceptance of internet-enabled smartphones presents an attractive opportunity for pavement condition evaluation. Mobile phones from manufacturers contain accelerometers to enhance functionality and updated their systems in new releases to be compatible with the current trends of maintenance planning. Modern smartphones are even equipped with 3- or 6-axis accelerometers which are used for a variety of applications within the phone's operating system. These accelerometers are sensitive enough to capture road roughness and defects data as transmitted through vehicle suspension, tires, and a mounting bracket. In general, the positive aspects of this approach include ease of installation, low cost, relatively large market for data gathering if released to the general public, and flexibility to be used with multiple agency-owned vehicles with a minimum of installation effort. Downsides include limited usefulness of generated data (due to limited correlation with existing pavement roughness indicators) and increased chance of operator error or vehicle modification issues if the operational plan includes public dissemination of the application. A system diagram of a typical smartphone-based data collection is shown in Figure 50.

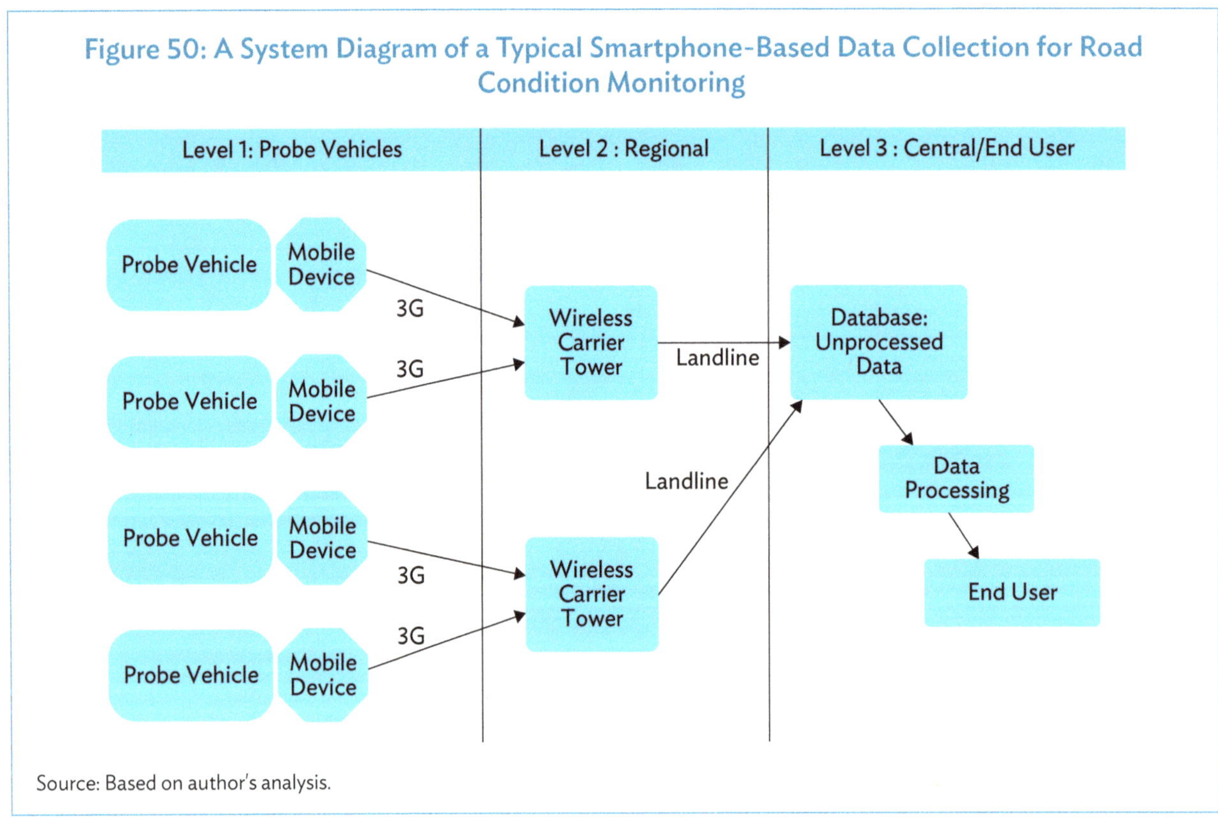

Figure 50: A System Diagram of a Typical Smartphone-Based Data Collection for Road Condition Monitoring

Source: Based on author's analysis.

The feasibility of using smartphone-based road condition monitoring methods will be discussed under the following subtopics: technical feasibility (system performance, integration/installation) and cost.

1. Technical feasibility: This describes the technical and hardware requirements for the implementation of the system. Smartphone-based assessment would be much easier to adopt compared to traditional methods which require additional equipment for data acquisition. For example, a smartphone-based system requires only that the smartphone screens face upward and the device heads point toward the front of the vehicle, while the x, y, and z axes of the accelerometer represent motion along the left–right, front–rear, and up–down directions of the vehicle, respectively. Other equipment such as GPS and video camera can also be placed on the dashboard to capture distress and other roadway data.[51]

2. Cost: A preliminary cost estimate allows the decision-maker or stakeholder to compare capabilities compared with the other alternatives. Compared with other techniques, smartphone-based applications are cost-effective since initial cost (purchased cost) and operational cost are both relatively low. Table 7 shows the comparison of the estimated cost for pavement roughness measurements smartphone-based method with that of the laser profiler (Class I), Bump Integrator (Class III) for a road network in Sri Lanka.[52]

Table 7: Comparison of Cost Values for Pavement Roughness Measurement Compared to the Laser Profiler (Class I), Bump Integrator (Class III) Illustrated from the Practice in Sri Lanka

Cost Item	Smartphone Application (Roadroid)	ROMDAS Bump Integrator	Laser Profilometer (HAWEYE 2000, ARRB system)
Operating Cost (SLRs/ per km) $1 = SLRs 300	125	2,000	2,250

SLRs = Sri Lanka rupees.
Source: Galagoda and Amarasinghe, 2019.

Various studies worldwide have already implemented the smartphone-based road condition assessments methods in the maintenance decision-making. Table 8 describes such applications which have been carried out recently.

[51] Sensor Event. 2012. *Android Developer Reference*. http://developer.android.com/reference/android/hardware/ SensorEvent.html
[52] D. Galagoda and N. Amarasinghe. 2019. Smartphone Applications for Pavement Roughness Computation of Sri Lankan Roadways. *Journal of the Eastern Asia Society for Transportation Studies, 13*, 2581–2601.

Table 8: Case Studies Conducted to Evaluate the Applicability of Road Condition Monitoring Using Smartphone-Based Approaches

Reference	Road Network/ Road Agency	Smartphone-based Data Collection	Performance Evaluation and Applicability on Maintenance Decision-Making
Fontanus et al.[a]	Rigid roads in Pennsylvania DOT, US	Roughness	Smartphone roughness slightly underestimated IRI compared with the inertial profilers, but the CV is almost same. Concluded that this system suitable for low volume roads, but for higher volume roads inertial profilers might be more appropriate.
NEXTRANS Project No. 098IY04[b]	Roadways in Illinois	Roughness	The repeatability of the roughness capture application was found to be acceptable for the intended application.
Roadroid[c]	Myanmar[d] Ministry of Construction (MOC) road network	Roughness and spatial data (distress via images/videos)	Established the applicability on suitable speeds depending on road type (ranging from concrete expressway to DBST and gravel roads in Myanmar.
	Afghanistan, Northern region roads	Roughness, GPS video data	Applicable for roads other than rough gravel roads and the very high level of deterioration (unmotorable condition)
	Swedish national roads	Roughness, GPS video data	Applicable for all type of tested roads.
StreetBump[e]	City of Boston roads	Potholes	Successfully identified different type of potholes with various severity and extend levels. Therefore, identified as an important emerging application. Boston aggregates the data across users to provide the city with real-time information to fix short-term problems and plan long-term investments.
Artis et al.[f]	Spain main roads	Pothole	The accuracy of the overall system is approximately 90%.
Mohan, Padmanabhan, and Ramjee[g]	Roads of Bangalore	Bumps and potholes, Heterogeneous traffic condition	For a speed of less than 25 km/h, the rate of the false negatives is 29% (well-oriented sensor) and 37% (virtually oriented). However, for a speed of more than 25 km/h, the rate of false negatives is 41% (well-oriented sensor) and 51% (virtually oriented). Concluded that the applicability is effective for such road network.
RoadLab[h]	Belarus Road Agencies	Roughness, GPS	Readily adopted RoadLab app to screen road surface conditions. Authorities are currently working on integrating it into their own road asset management database and the system of the Traffic and Road Safety Coordination Center.

continued on next page

Table 8 continued

Reference	Road Network/ Road Agency	Smartphone-based Data Collection	Performance Evaluation and Applicability on Maintenance Decision-Making
Bump Recorder[i]	**Private Sector** TOA Road Corporation.	Roughness, GPS	Introducing BumpRecorder to many countries on JICA KCCP (The Knowledge Co-Creation Program; Japanese International Cooperation Agency Educational Program)
	National agency - MLIT (Ministry of Land, Infrastructure, Transport and Tourism) **Local government** - Aizu-wakamatsu city, Tokyo Kita ward, Katsushika ward, Shizuoka prefecture and city **Public authorities** JICE (Japan Institute of Country-ology and Engineering) NILIM (National Institute for Land and Infrastructure Management) NIED (National Research Institute for Earth Science and Disaster Resilience) **Academia** Tokyo University,Kyoto University		

CV = computer vision, DBST = double bitumen surface treatment, GPS = global positioning system, IRI = international roughness index, km/h = kilometer per hour.

[a] M. Fortunatus et al. 2018. Use of a Smart Phone based Application to Measure Roughness of Polyurethane Stabilized Concrete Pavement. *Civil Engineering Research Journal.*

[b] B. William and S. Islam. 2014. *Integration of Smart-Phone-Based Pavement Roughness Data Collection Tool with Asset Management* [ac]

[c] Roadroid. 2022. Roadroid. https://www.roadroid.com/.

[d] Effective 1 February 2021, ADB placed a temporary hold on sovereign project disbursements and new contracts in Myanmar.

[e] Street Bump. 2020. Where's Street Bump being used? http://www.streetbump.org/about System. USDOT Region V Regional University Transportation Center.

[f] M. Artis et al. 2011. Real Time Pothole Detection Using Android Smartphones with Accelerometers. 1–6.

[g] P. Mohan, V. Padmanabhan, and R. Ramjee. 2008. Nericell: using mobile smartphones for rich monitoring of road and traffic conditions. *6th ACM conference on Embedded network sensor systems,* ser. SenSys '08, 357–358.

[h] Wang, W. W. n.d. *BIG Revamping Road Condition and Safety Monitoring with Smartphones.* https://olc.worldbank.org/system/files/WBG_BD_CS_Road_Monitoring.pdf

[i] Koichi, Y. 2016. BumpRecorder. https://www.bumprecorder.com/document/161021_BumpRecorder_workshop.pdf

Limitations of the Technology and Challenges in Implementation

Smartphone-based apps provide the opportunity to deliver continuous monitoring and thus generate big data. As described in previous chapters, smartphone-based data is relatively cheaper than all other conventional methods as the emerging technology on smartphones can generate accurate data in large quantities by using crowdsourcing techniques.

However, there are limitations ahead in implementing such systems instead of typical road condition monitoring tools. The limitations can be categorized into several types such as limitations in data collection, information management, and application development. The limitations are mostly applicable to individual distress data collection. In the case of the reporting road roughness over a road segment, some of these issues will have less significance.

Inconsistency in vibration pattern data is one of the predominant technological issues in the data processing stage. A given distress such as pothole or any other road anomaly may not necessarily give the same pattern during each drive over it. The sensor signals can vary according to the particular type of mobile device, and various device characteristics, such as the hardware and operating system, can impact the signal.

Moreover, there can be network overloading and delay when the sensor data is uploading to the server at back end. If a large amount of data is needed to be sent over the network, it may lead to network congestion leading to delay or loss of data. Therefore, the application must keep the network usage to minimal. Scalability of the supporting IT infrastructure can be a key concern when collecting data from mobile devices, especially if it's expanded to get data from the public who will use the app.

GPS provides the latitude and longitude values of a location. It is used to detect the location of road defects to users. It may also be possible to miss some GPS data in rural areas due to insufficient signal strength especially in developing countries, which can lead to a higher margin of error. This is also more relevant to distress identification, since roughness is reported for road segments, loss of intermittent GPS locations may not be a major issue.

There can be a privacy issue when the smartphone application requires the location of the device to detect the location of road anomaly and can be retrieved by user; hence it may lead to a privacy breach.

The data quality technology in mobile devices can limit the quality of the information gathered from the devices, compromising precision and limiting the use of the data for its intended purposes. The sensitivity of components such as accelerometers or GPS capabilities differs according to the device.

User behavior can also influence what the device finds. For example, a driver might consciously slow down or drive around potholes, which clearly limits the potential success of the app in finding potholes. Each behavior would result in different findings on the exact environmental condition.[53]

User incentives and expectations management is a key challenge in using a mobile app to facilitate data collection. If the app is too costly or there are no incentives for using it, the amount of data may

---

53   D. Bisconsini et al. 2018. Pavement Roughness Evaluation with Smartphone. *International Journal of Science and Engineering Investigations, 7,* 43–52.

not be large enough. Further, users may anticipate that by driving with the app on in their car, potholes would be identified, and the city would use that information to fix the potholes. There would be a direct user benefit because potholes would more likely be identified and fixed. To accommodate those expectations, organizations need to provide feedback in different situations.

Anticipated improvements in the future

Efforts have been made to implement various methods which detect road surface anomalies using data from smartphone sensors. Regarding the performance of the relevant algorithms, it is difficult to compare the accuracy and performance of various approaches due to restricted availability of the reported algorithms and datasets. For many threshold-based approaches, the manner in which the threshold was set remains unclear. In addition, methods that use supervised or unsupervised learning methods need large amounts of data to train their detection model. Therefore, a hybrid approach, which would be able to continually detect and distinguish various road surface anomalies, using real-time data streaming from smartphone sensors and other geographic data, should be developed.

Sensor data values should be smoothed and reoriented to allow smartphone users more freedom, as well as to increase the accuracy of detection. The ideal approach should be self-adapting and self-learning; it would be able to reconcile itself to any platform, the dynamic behavior of different vehicles, as well as various road surface conditions. For that a free, cross-platform smartphone application must be developed. This application would be fully automatic and not require any user interaction, as well as allow the driver to maintain focus on the road.[54]

In addition, the power consumption of the proposed application should be minimized, perhaps by reducing the usage of GPS sensors. Specific techniques, such as robust sensor calibration readings, should also be investigated because the amount of sensor noise and accessibility of the sensor data vary in different smartphones. Due to the increase in the number of smartphone holders, continuous monitoring and reporting of road surface anomaly events are achievable by the public. In fact, the effectiveness of road monitoring systems could be substantially improved with effective data-crowdsourcing techniques. Smartphones generally are embedded by various wireless interfaces, such as Bluetooth, Wi-Fi, and cellular networks, making them ideal for crowdsourcing purposes and pushing notifications to road users.[75]

---

[54]   S. Sattar, S. Li, and M. Champmon. 2018. Road Surface Monitoring Using Smartphone Sensors: A Review. *Sensors*.

Moreover, vehicular network technologies, including V2V (vehicle-to-vehicle) and C-V2V (cellular-V2V) communication technologies, are emerging and can play a critical role in transportation management and specifically on driving safety.[55] With more vehicles being connected to the internet and to each other, there is potential for enhancing the processes of:

1. Data collection by obtaining the most updated information regarding the road surface conditions from all connected vehicles.

2. Data exchange among road users by notifying drivers approaching road surface anomalies detected by other road users beforehand through V2V and C-V2V technologies. However, these technologies are still under investigation and are not widely deployed on vehicles due to the absence of a road infrastructure.[56]

---

[55] C. Silva et al. 2017. A survey on infrastructure-based vehicular networks. *Mobile Transportation System*.

[56] B. Masini, A. Bazzi, and A. Zanella. 2018. A Survey on the Roadmap to Mandate on Board Connectivity and Enable V2V-Based Vehicular Sensor Networks. *Sensors*.

IX. Moving Forward—Integrating Different Techniques in Road Condition Monitoring

This chapter discusses how frontier technologies can improve efficiency in pavement condition monitoring programs. It clearly outlines its scope based on the technology available as well as considering the financial implications.

The scope of frontier technologies would be largely limited to network-level assessment and can be mainly applied to identify the overall condition of the pavement (categorical analysis) or roughness. The traditional methods still need to be applied for road condition quality assurance investigation and project-level assessment where more detailed information is needed. Also, it is emphasized that in certain situations where traditionally used equipment are not available, satellite imagery or smartphone-based data collection methods can play a major role.

The cost comparison reveals that smartphone-based methods are very cost-effective compared to traditional methods when it comes to roughness determination. While acquiring high-resolution satellite imagery is typically more expensive than traditional methods, a robust machine learning model could help mitigate these costs. Leveraging AI technology may reduce the frequency of satellite imagery acquisitions compared to conventional approaches.

The way forward in integrating all these technologies is also discussed in this chapter. It is recommended that frontier technologies be used to monitor overall condition and roughness of the pavement at network level. The advantages it offers for periodic monitoring is also elaborated. It is also recommended that the use of satellite imagery should be done after assessing the cost-benefit analysis outcomes for the selected road network.

Comparison of Pavement Condition Monitoring Techniques: Traditional versus Frontier Techniques

The traditional techniques of pavement condition monitoring mainly involve on-site surveys either done manually or with automated techniques. Manual techniques mainly involve distress data collection, skid resistance measurement, and structural strength assessment. Automated techniques employ laser profiling, LiDAR, video image processing, and vehicle response-based measurements. Both manual and automated methods are able to monitor all the basic pavement performance metrics such as frictional performance, structural condition, distress condition, and overall pavement condition. This can easily be adopted for network level as well as project-level pavement condition monitoring.

Frontier technologies such as smartphone or satellite image-based methods are mostly adopted for overall pavement condition monitoring and are not yet developed to identify the other pavement performance metrics such as skid resistance, structural condition with any level of accuracy, and practicality when implemented on the field. The best practices of using frontier technologies show that it has potential in network-level condition monitoring especially with respect to categorization of pavement condition ratings and network-level roughness assessment.

The techniques are more advantageous as they can be used to monitor condition in remote locations which would have difficulty in accessing if traditional equipment is to be deployed. Also, there is significant cost savings due to the lower mobilization cost to commence such work in remote areas.

Pavement condition monitoring and data processing can be carried out in a centralized location which benefits road agencies in rural areas with limited resources.

There is also potential for reduced time for condition monitoring and evaluation which could be useful in feasibility study evaluations for road improvement projects.

Frontier technologies also offer a way forward in improving continuous pavement monitoring, especially for multi year asset management programs where the required equipment will not be available as needed by the road agency.

Frontier technologies also improve the transparency in the condition monitoring results as it can be verified using multiple sources.

Summary of Financial and Technical Resource Requirements

An accurate cost comparison of these techniques would depend on factor such as density of the road network, and terrain and land use. A preliminary assessment can be done using the cost values published in case studies.

Based on a case study conducted in New Castle County, the total cost of using satellite imagery-based techniques varies from approximately \$850 per year per square kilometers (km^2)[57] (equivalent to \$815 per year per km based on the road network) to \$15 per year km^2 depending on whether the images are high resolution (SAR images) and the monitoring is done in house or outsourced.

A study in African countries also provides values ranging \$30–\$90 per km^2 for satellite image-based assessment using Airbus satellite imagery and manual categorization. The corresponding cost for traditional assessment was in the range of \$12–\$39 per km^2. However, the processing time for satellite imagery-based technique was 2.5 times higher than the traditional method.

---

[57]   The life-cycle cost given in the paper by Li et al. (2017) has been converted to an annual cost by dividing the total present value of the project cost by the analysis period of 20 years.

A comparison of satellite image acquisition costs published by satellite image providers also suggest that prices vary from \$11 to \$40 per km^2. Comparison of per km value would not be fair since this largely depends on the road density and network distribution in the area.

The cost of conducting traditional pavement roughness monitoring methods such as using the bump integrator, or the laser profiler is \$8–\$9 per km based on results available in Sri Lanka. The cost of smartphone-based roughness measurement would be in the range of \$0.5–\$1 per km depending on the network length and the road density in the area. This includes the software license fee, and transport and data processing fees (if any, since typically data processing is carried out by the software itself). The licenses are usually given for a period and the cost per km depends on the total number of km in the respective road agencies network.

To demonstrate the cost comparison between the two technologies, an illustrative example is presented based on a characteristics of a provincial road network in Sri Lanka. This shows the per km cost for satellite-image-based method and smartphone-based method to be \$16.30 per km and \$0.68 per km (\$1 = SLRs300), respectively. These values are within the range of values assumed in the previous paragraph of this section.

Table 9: Cost Comparison for Smartphone-Based and Satellite-Image-Based Monitoring for Illustrative Example

Road length (km)	3500 km		
Area (km^2)	2763 km^2		
Type of roads	Provincial roads in Western province Sri Lanka		
Satellite imagery		Smartphone	
Procurement cost			
Project duration	2 months	Project duration	100 km/day 35 days
Satellite imagery purchase cost (\$20 per km^2)*	\$0–\$55.260 SLRs0–SLRs16,332,995	App purchase	SLRs100,000 (Roadroid)
GIS analyst	SLRs200,000	Operating cost (includes vehicle, fuel, and driver) SLRs125/km	SLRs437,500
Data scientist (1.5 man months)	SLRs450,000	Smartphone device	SLRs50,000
Computer rental SLRs27,778 per month**	SLRs55,556	Laptop/desktop computer rental***	SLRs7,000
Highway engineer (0.5 man months)	SLRs60,000	Highway engineer / technical officer (1 man month)	SLRs120,000
Internet charges (SLRs15,000 per month)	SLRs45,000		
Total	SLRs17,143,551	Total	SLRs714,500

*Satellite imagery cost can be further reduced by identifying the most relevant sections (e.g., purchase strip maps of required road sections).

**Overhead cost for a computer compatible for satellite image processing (SLRs1 million) applied for the project duration.

***Overhead cost for laptop/desktop computer for smartphone roughness data analysis (SLRs250,000) applied for 1 month of project duration.

Irrespective of whether traditional techniques or frontier technologies are used, the capacity building of road agencies staff is required in order to operate the equipment, obtain and process the data. Traditional roughness measurement methods require purchase of expensive equipment such as the laser profilometer, which costs around $50,000, and the bump integrator, which costs around $15,000. The other resource requirements are not a major concern as high-end computers, smartphones, and data storage facilities are usually available in road agencies. Alternatively, road agencies can use cloud storage (i.e., government cloud storage facilities, to reduce the cost if necessary). If this is required for a relatively smaller network, it would be prudent to outsource it since the investment in in-house equipment and staff will increase significantly the overhead cost of road monitoring. A cost-benefit analysis must be done on a case-by-case basis to decide the best approach.

However, satellite image-based methods still remain the costliest option out of all available methods. The only advantage being that it would expedite the processing time, especially if it's done on a large, sparse road network. The cost of satellite imagery acquisition could be reduced if there is mechanism to share the acquisition cost with other state agencies, such as irrigation, agriculture, or disaster management, which may benefit from those images. In terms of cost, there is significant gain in using smartphone-based methods for roughness data collection and, considering its acceptable accuracy levels, it can be adopted for network-level road condition monitoring.

Discussion on Integrating Different Techniques in Road Condition Assessment

The way forward in adopting the frontier technologies largely lies in the integration of these tools in asset management. The following section discusses one such approach where the strengths of each technology can be adopted to improve the road condition monitoring and can be used for planned maintenance program (or when roads are needed to be selected for a rehabilitation project)

First, it is important to understand the role of each technology, in terms of the type of metrics of parameters representing the road condition. Available evidence suggests that frontier technologies are best used to determine the overall pavement condition (aggregate level condition monitoring rather than monitoring individual distress condition), especially at the network level. The next question is, what are the metrics that can be used to monitor overall pavement condition? The two most common types of metrics are the pavement condition rating (qualitative or quantitative) or a parameter such as pavement roughness.

Satellite-based image processing is viable for preliminary screening of roads based on the overall condition. The research carried out thus far proves that it is successful in classifying roads up to three levels with high level of accuracy. Therefore, the road agency can leverage on this tool to categorize road network under, for example, as "good," "satisfactory," and "poor" conditions that would facilitate preliminary level screening of the network. This is advantageous especially for road agencies with road networks spanned over large areas of land that would otherwise need on-site investigations to assess condition.

The roads that are screened as "good" from the first stage of the analysis can be excluded from the second stage. The main reason is that they do not need to be considered in the rehabilitation or planned maintenance program. The condition monitoring for remaining categories of roads would need to have a more quantitative parameter that would facilitate computation of maintenance cost, remaining service life determination, etc., that would be inputs required for the road agency's decision-making. Thus, pavement roughness evaluation can be selected as an effective metric to assess the remaining road sections categorized as 'satisfactory and poor'. This can be carried out with sufficient accuracy using smartphone-based roughness assessment methods. Several research findings have validated that smartphone-based techniques provide sufficient and accurate roughness data for maintenance related decision-making. As explained earlier, this has been proven as a cost-effective monitoring tool.

With roughness data as input to the asset management system, roads that needs to be included in the planned maintenance program or rehabilitation project can be identified. If additional data are required for roads selected for the maintenance work, this can be obtained by the traditional manual surveys or the image processing-based techniques that can identify distresses. While the manual surveys require minimum technological input but more technical staff and time to conduct, the image processing-based techniques can automate the data collection process thus minimizing the time required. However, this requires additional resources in terms of trained technical staff, vehicles and equipment to capture and process the images. Therefore, it is prudent to minimize the quantity of roads that require this detailed investigation.

X. Summary and Conclusion

The study first introduced the fundamental concepts of pavement condition monitoring—the performance metrics used by road agencies and what they represent in terms of a road's functional and structural performance. These include, among others, pavement friction, distress condition, structural condition, roughness. It also discusses the importance of these metrics, internationally accepted standards that are used in maintenance decision-making. The equipment and other technologies that are traditionally used to measure these performance metrics are also discussed. Most performance measurement methods have advanced from manual methods to automated techniques with the advent of technologies such as video processing, vehicle response detection, LiDAR, and laser profiling. This has improved the accuracy and the consistency in the condition monitoring as well as expedited the surveys.

One of the main limitations of these traditional methods is the availability of the equipment especially in rural road networks in developing countries. The high capital cost and the hands-on training required to operate makes them less accessible to provincial or local agencies. Some equipment, such as the laser profilometer cannot navigate in the narrow and highly deteriorated road segments. Thus, the reliance of such methods inhibits quantitative assessment of the road condition that would facilitate objective maintenance decision-making for the road agencies in these areas.

The adoption of frontier technologies such smartphone-based methods and satellite image-based methods have the potential to resolve some of these issues that are inherent with traditional technologies.

In particular, smartphone-based technologies have proven to offer a low-cost accurate method to measure pavement roughness on roads. This requires limited training for technical staff involved. It also doesn't require dedicated vehicles, which will maximize the resource usage in the road agency. The data storage and processing can be done remotely on a cloud platform, which would allow a database to be maintained and pavement data analytics to be done at a centralized institution. This will eliminate the issues arising due to lack of technical competence in local road agency staff, for example. Smartphone-based approaches can also be adopted for distress identification, but this would require additional image processing software and better video image capturing and storage devices than what is available in a typical smartphone. The calibration and training requirement in the machine learning algorithm used would also be more specific to the local agency.

Satellite imagery-based technology offers a completely remote platform for pavement condition monitoring. This can be implemented with minimal involvement of the local road agency, provided the condition assessment done through satellite image processing is validated through ground-truthing surveys. This has the potential to cover a large road network with minimum time and also would be easier to carry out periodic surveys for multi-year asset management systems. The major limitation is the type of pavement performance metric that can be accurately monitored using this technique is limited to the overall pavement condition, typically presented in a three-level categorization (up to five levels with lower accuracy). The survey cost has also appeared to be higher than smartphone-based methods as well as manual methods considering the mainly the high cost of satellite image acquisition. However, with an efficient and reliable machine learning model and AI technology, the survey, using satellite imagery, can be done at a lower frequency compared to the traditional methods, resulting to a more cost-effective road condition monitoring. Several state agencies can also consider sharing the cost, while the data acquisition and analysis process can be outsourced to a third-party organization. With understanding the outputs will be shared among the stakeholder organizations, the satellite image-based techniques can be a cost-effective road condition monitoring as well. It is also important to note that the analytical tools described in this guidebook are not intended to replace traditional methods. Rather, they can serve as complementary approaches for a more comprehensive, timely, accurate, and flexible method of monitoring..

In the long term, there is an opportunity to integrate these technologies to improve the overall efficiency of the road network condition monitoring. Frontier technologies can be adopted successfully for preliminary level screening of the road networks for both network-level and project-level pavement asset management, while the traditional methods can be applied to the selected roads for detailed investigation that would provide more outputs necessary for effective maintenance management of the road network.

Appendix

These sections provide a detailed discussion of conventional approaches for monitoring road quality, with specific focus on road pavement conditions.

TRADITIONAL TECHNIQUES

Manual and Visual Techniques

Manual and visual data collection methods refer to road quality assessments that require a human involvement on-site to record the relevant characteristics of the pavement condition metric(s) being assessed. If these characteristics are obtained either using images, sensor, or other equipment that doesn't require direct human intervention, it is considered as automated data collection, which will be discussed in the next section.

Manual methods can be used to measure most of the pavement condition metrics mentioned in the previous section, either directly or indirectly.

The following pavement condition metrics can be assessed manually:

1. *Roughness*. Roughness can be measured using the precision profiling device or estimated based on the present serviceability rating (PSR) value, which the subjective rating is given by the evaluator.
2. *Distress condition*. Walk-through surveys or windshield surveys where experienced inspectors identify the existing distresses on pavement segments and rate the pavements. In the manual data collection, data are recorded on paper or handheld data loggers, and shall be analyzed either on paper or with a computer after being transferred into a database. The distress data collection is time-consuming with this method and typically distress type, severity, and density are collected to evaluate pavement condition. When manual technique is adopted at network-level evaluation, inspection units may range in lengths from 20 to 100 meters and are selected at random or through a defined sampling procedure, since covering the entire network is not viable. For example, in the American Society for Testing Materials (ASTM) standard for pavement condition index (PCI) calculation, the sample size recommended is calculated as follows:

   - The minimum number of sample units (n) is determined based on the following equation:

$$n = Ns^2/((e^2/4) (N-1) + s2)$$

   e = acceptable error in estimating the section PCI; generally, e = ±5 PCI points;
   s = standard deviation of the PCI from one sample unit to another within the section. In the initial inspection s is assumed to be 10 for asphalt concrete (AC) pavements and 15 for portland cement concrete (PCC) pavements
   N = total number of sample units in the section

- If obtaining the 95% confidence level is critical, the adequacy of the number of sample units surveyed must be confirmed. For example, if there are 200 road segments to be surveyed, the sample size would be 15. Therefore, this method may not represent the condition of the entire road link sufficiently, especially if the condition varies.

In a typical windshield survey, the survey is done from a vehicle traveling at a speed of about 10 to 25 kilometers per hour (km/h). Manual technique is prone to inconsistency on field assessment of the pavement condition, especially when there is lack of trained technical staff. To alleviate this issue to an extent, especially when making assessment on the overall pavement condition, detailed guidelines such as Pavement Surface Evaluation Rating (PASER) Manual is available to assist the technical staff conducting the evaluation. Also, most road agencies have published the distress condition guidelines, which provides a comprehensive methodology on how to measure each distress type.

3. *Pavement structural condition.* Pavement structural condition assessment methods generally fall under manual techniques. Methods such as Dynamic Cone Penetrometer, California Bearing Ratio tests which assess the strength of the pavement layers, including the sub-grade require direct intervention of the technical staff. Even the Falling Weight Deflectometer or ground penetrating radar (GPR) test if it's done at isolated segments tests can be considered as manual techniques.

4. *Pavement friction.* The British Pendulum Tester/Dynamic Friction Test is a friction test that can be done manually. In addition, macro-texture measuring methods such as the sand patch test falls within the manual technique.

5. *Pavement marking.* Visual inspection of pavement marking evaluation and use of mobile and handheld retroreflectometers are manual technique used to evaluate the quality of a marking.

Automated Data Collection Technologies

Automated data collection refers to pavement data collection carried out with the assistance of sensors, video, images, scanners, radar, laser, etc. The data are still collected in the field and not remotely, and therefore the equipment required to acquire the data via these methods should be fixed to the survey vehicle which needs to travel along the road. The difference with the manual techniques discussed above is that it minimizes human involvement and automates the monitoring and process of measuring the relevant pavement condition characteristics. Furthermore, additional analytics are involved to transform the data gathered through various techniques to know parameters that we used to assess the pavement condition using a particular metric. For example, images of road cracks can be used to derive information relevant to crack type, severity, and extent using image processing techniques and machine learning.[1] The relevant analytics is often carried out by proprietary software available with the equipment used for the survey.

---

[1] N. Attoh-Okine and O. Adarkwa. 2013. *Pavement Condition Surveys—Overview of Current Practices.* Newark: Delaware Center for Transportation, University of Delaware.

Laser Profiling

Laser profilers are used to minimize the errors and standardize the survey process by recording, reducing, processing, and storing pavement data using several equipment attached to a multifunction vehicle. This technology holds great promise in the area of automated high-speed distress data and several types of systems have been developed by the researchers.

Typically, a network survey vehicle is equipped with a digital laser profiler (DLP) which has the capability of estimating various data types in higher accuracy and precise levels. However, most of the equipment can collect data for 3.6 meters of width at once, which reduces survey time and costs due to ability to collect data at highway speed (usually up to 100 km/h). This type of profilers has advantages over other techniques, such as getting results that are independent of vehicle type, measurements that are possible on all sealed surfaces, and data linked to chainage and global positioning system (GPS) coordinates.

These laser measurements produce outputs such as roughness, rutting, longitudinal profile, faulting, transverse profile, raveling, and macro texture.[2] ARRB (2000) illustrates the overview of the pavement condition metrics which can be measured by laser profilers (Figure A.1).

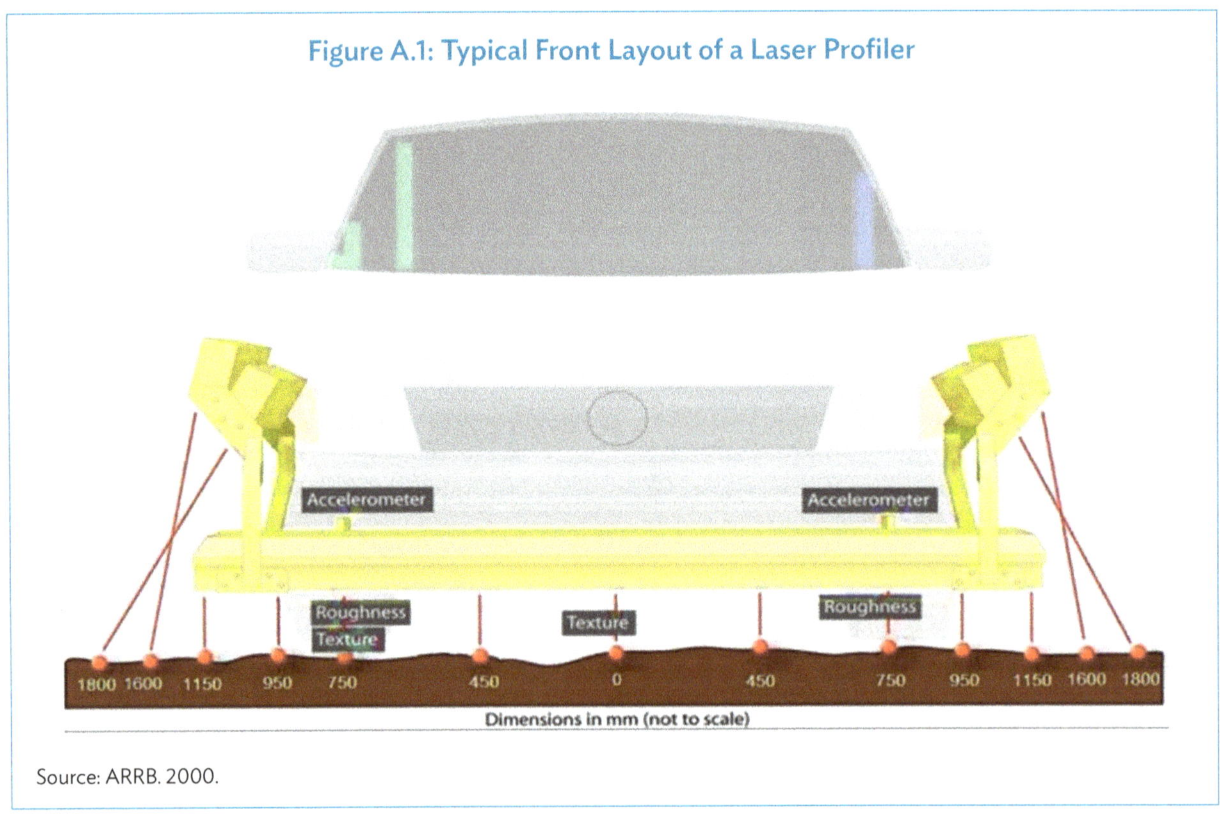

Figure A.1: Typical Front Layout of a Laser Profiler

Source: ARRB. 2000.

---

2   ARRB. 2000. *ARRB System*. Retrieved from Haweye 2000: https://arrbsystems.com/fact-sheet/hawkeye-2000/

Table A.1: Overview of Pavement Condition Metrics Measured Through Laser Profilers

Pavement Condition Metric Assessed	Attributes Measured for Each Metric	Laser Profiler Type(s) / Road Agency
Roughness	Longitudinal profile	ROMDAS Laser Profiler[3] ARAN System[4] PaveProf V2.0[5] LCMS-2[6] Laser Profiler NTUA[7]
	Profilograph Index (PI), RQI, Half Car Ride Index (HRI), Ride Number (RN)	PaveProf V2.0
Rutting	Rut type, width (single or double), depth, cross section area and percentage of deformation	ARAN System PaveProf V2.0 LCMS-2
Macro texture	Estimated texture depth or equivalent sand patch texture depth	Haweye-2000 ROMDAS Laser Profiler ARAN System PaveProf V2.0 LCMS-2
Faulting	Elevation difference	Haweye-2000

Source: Authors

Light Detection and Ranging (LiDAR)

LiDAR is a remote sensing method that uses pulsed lasers to measure variable distances to the earth. These light pulses are combined with other data which have been collected from the airborne system and generate precise, three-dimensional information regarding the shape of the earth's surface.[8]

This technique is adopted to identify pavement distress using either low-altitude unmanned aerial vehicles or wheel-based mobile mapping systems to collect image data. Later, the classification of such data is done using classification techniques such as random forest classification. In brief, the data collection has several steps as illustrated in Figure A.2. First, road surface condition extraction is done by using automatically extracted 3D point clouds with the aid of vehicle trajectory elevation information. In the second step, classification and clustering of 3D points is done for the road surface as well as below it, which correspond to the crack depth and pothole quantification. Finally, each individual distress location is identified as clusters by boundary tracing and quantitative analysis, which can be visualized by back projecting the detected boundary on the corresponding red, green, and blue (RGB) imagery.

---

[3]   ROMDAS Data Collection Ltd. 2011. ROMDAS System. Retrieved from Laser Profilometer: https://romdas.com/romdas-laser-profiler.html

[4]   M. Sršen. 2002. Automatic road analyzer- ARAN.

[5]   PaveTesting® Ltd. 2011. Profiling & Digital Imaging. United Kingdom: PaveTesting® Ltd.

[6]   Pavemetrics. n.d. *Pavemetrics*. Retrieved from Pavemetrics® Laser Crack Measurement System (LCMS®-2): https://www.pavemetrics.com/applications/road-inspection/lcms2-en/lcms-2-roughness-iri/

[7]   A. Loizos and C. Plati. 2008. Evoluational Process of Pavement Roughness Evaluation Benifiting from Sensor Technology. *International Journal on Smart Sensing and Intelligent Systems*, 370–387.

[8]   NOS. 2021. *What is lidar?* Retrieved from National Oceanic and Atmospheric Administration, US Department of Commerce: https://oceanservice.noaa.gov/facts/lidar.html

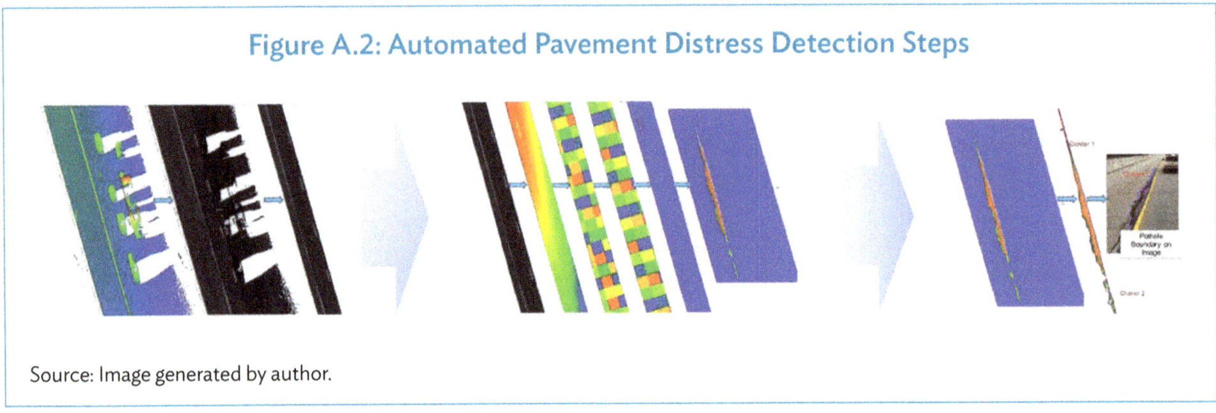

Figure A.2: Automated Pavement Distress Detection Steps

Source: Image generated by author.

The LiDAR method is often used to detect potholes. Potholes have different surface areas and depth, which is then expressed in terms of severity levels and can be visualized through these RGB images. Identifying low-severity potholes requires high-grade mobile mapping systems, while medium- to high-severity potholes can be easily detected from typical systems. Usually, the elevation of potholes is lower than that of the pavement surface Also, asphalt mixture in pothole areas is usually worn away, exposing the roadbed and later on getting embedded with foreign materials. Because of this, the spectral features of surfaces with potholes are different in compared with the other pavement areas. Moreover, potholes are either circular or oval-shaped, making them easier to capture and screen in LiDAR systems.

Generally, several distress types can be seen at a particular road segment, such as cracks, rut, and potholes. Moreover, other distress patterns can be captured through those special features in either color or elevation profiles as in Figure A.3.[9] In addition to detecting pavement distress types, LiDAR systems have been adopted to identify several roadway geometries, such as cross slope, super elevation, etc.[10] Moreover, Table A.2 illustrates the overview of the pavement condition metrics which can be measured by LiDAR.

---

[9]   Z. Li et al. 2019. Identifying Asphalt Pavement Distress Using UAV LiDAR Point Cloud Data and Random Forest Classification. *International Journal of Geo-Information, 8*(39).

[10]   A. Famili. 2020. *Pavement Surface Evaluation Using Mobile Terrestrial LiDAR Scanning Systems.* Graduate School of Clemson University.

Figure A.3: Distress Patterns Captured Through Color or Elevation Profiles

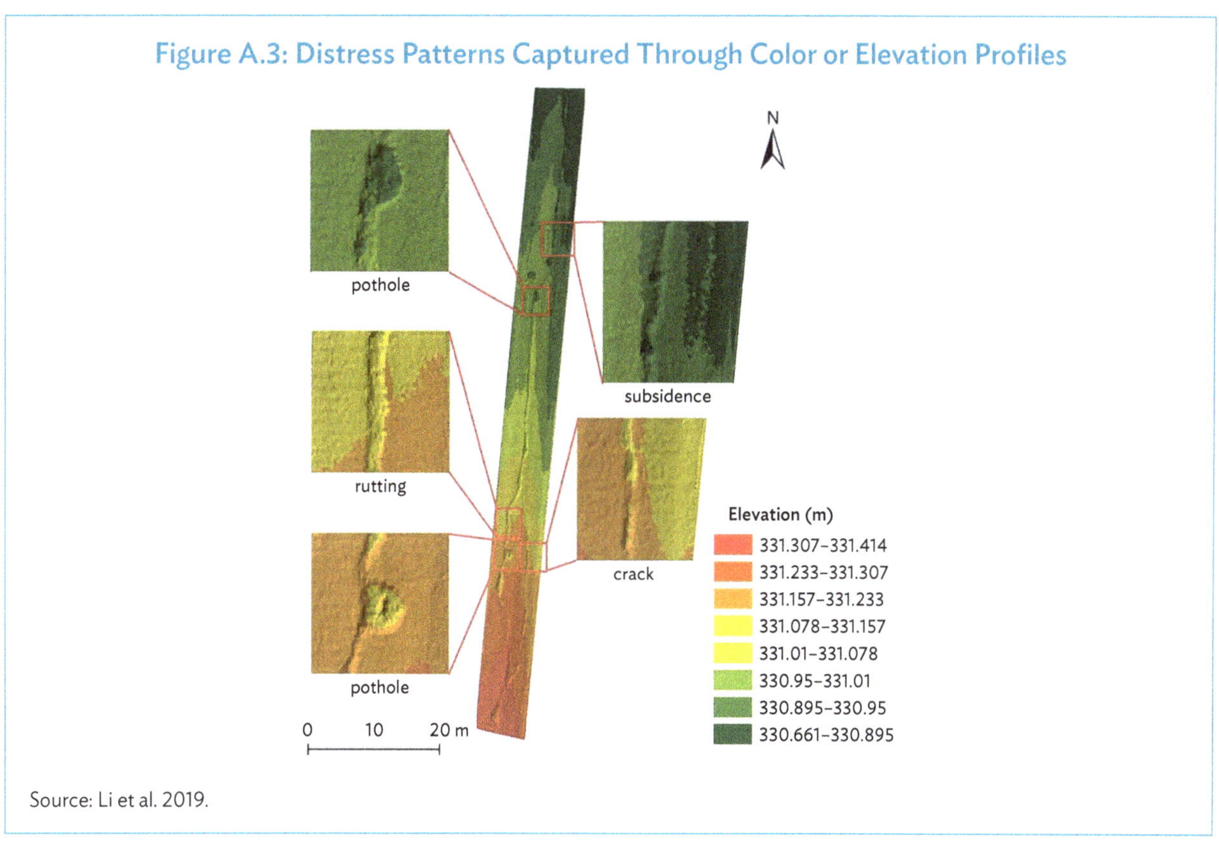

Source: Li et al. 2019.

Table A.2: Overview of Pavement Condition Metrics Measured Through LiDAR

Pavement Condition Metrics Assessed	Attributes Measured for Each Metric	Case Study/Road Agency
Distress	Identify and estimate pothole volume to calculate the amount of fill material needed	Hsinchu County in northwestern Taipei,China[a]
	Location, severity, cause, and a quantified estimate of the spatial and volumetric characteristics of potholes	United States[b]
	Spatial, volumetric characteristics and severity of potholes, rutting, transverse cracks and longitudinal cracks	Shihezi City in the Xinjiang Uygur Autonomous Region of the People's Republic of China[c]
	Pavement crack width	Georgia[d]
Roughness	Dynamic IRI, geometric IRI computed on a digital elevation model (DEM)	Italy[e]
Road surface extraction	Location and approximated direction of local road patch	Xiamen, People's Republic of China[f]

continued on next page

Pavement Condition Metrics Assessed	Attributes Measured for Each Metric	Case Study/Road Agency
Road cross section	Cross slope	Georgia[g]
	Number of lanes, width of the roadway, width of the shoulders, width of the lanes, super elevation	Spain[h]

IRI = international roughness index, LiDAR = light detection and ranging.

[a] K. Chang, J. Chang, and J. Liu. 2005. Detection of pavement distresses using 3D laser scanning technology. 2005 ASCE International Conference on Computing in Civil Engineering.

[b] R. Ravi, D. Bullock, and A. Habib. 2020. HIGHWAY AND AIRPORT RUNWAY PAVEMENT INSPECTION USING MOBILE LiDAR. *The International Archives of the Photogrammetry, Remote Sensing and Spatial Information Sciences*, 349–356.

[c] Z. Li et al. 2019. Identifying Asphalt Pavement Distress Using UAV LiDAR Point Cloud Data and Random Forest Classification. *International Journal of Geo-Information, 8*(39).

[d] Y. Tsai, C. Jiang, and Z. Wang. 2012. Pavement Crack Detection Using High-. *7th RILEM International Conference on Cracking* in Pavements, 169–178.

[e] M. Blasiis et al. 2021. Assessing of the Road Pavement Roughness by Means of LiDAR Technology. *coatings*.

[f] H. Wang et al. 2012. Automatic road extraction from mobile laser scanning data. *Computer Vision in Remote Sensing*, 136–139. 2012 International Conference on IEEE.

[d] Y. Tsai et al. 2013. Mobile Cross-Slope Measurement Method Using Lidar Technology. *Transportation Research Record Journal of the Transportation Research Board*, 53–59.

[h] B. Holgado et al. 2017. Automatic Inventory of Road Cross-Sections from Mobile Laser Scanning System. *Computer-Aided Civil and Infrastructure Engineering, 32*(1), 3–17.

Source: Authors.

Video and Image Processing

Video or image processing techniques involve a camera-based solution that leverages computer vision algorithms or neural network methods to evaluate pavement condition-related metrics, such as potholes, cracking, aging, depressions, texture, etc. Advancements in the quality of cameras, data storage, machine learning tools, and image processing technology have made this approach technically feasible and cost-effective for pavement condition monitoring.

This type of pavement condition evaluation consists of three basic components: (i) data acquisition, (ii) storage, and (iii) image processing. Generally, data acquisition is done by capturing the pavement surface images while traveling at normal speeds, while image processing is typically carried out in the following steps: image digitizing and deburring, image segmentation, clustering, feature extraction, and cluster classification.[11] Image digitizing and deburring is the conversion of the raw images or videos into specific resolution levels and set in arrays to segmentation based on the features. Image segmentation is done to identify the location of a single distress in the given image. In case there are multiple distresses present, it has to rely on the concept of object detection. There are various segmentation techniques to evaluate pavement conditions such as histogram thresholding,[12] one- and two-dimensional entropy-based approaches,[13] edge detectors,[14] and vertical region segmentation.[15]

---

[11] A. Adolfo et al. 1992. Low-Cost Video Image Processing System for Evaluating Pavement Surface Distress. *TRANSPORTATION RESEARCH RECORD*, 63–73.

[12] S. Ritchie. 1990. Digital Imaging Concepts and Applications in Pavement Management. *Journal of Transportation Engineering, 116*(3), 287–298.

[13] A. Abutaleb. 1989. Automatic Thresholding of Gray-Level Pictures Using Two-Dimensional Entropy. *Computer Vision, Graphics and Image Processing, 47*, 22–32.

[14] M. James. 1988. Paflem Recognition. John Wiley & Sons.

[15] J. Acosta. 1991. Implementation of the Video Image Processing Technique for Evaluating Pavement Surface Distress. *MS Thesis, Case Western Reserve University*.

Clustering is then done by linking all adjacent foreground and connective pixels so that the cluster is surrounded only by background pixels. This is then followed by feature extraction, or identifying the features characteristic of a particular type of distress following a set of rules. Finally, the clusters are classified into a specific group or feature and assigned accordingly.

Figure A.4 illustrates the process of image processing and Figure A.5 shows the raw images taken into the analysis and the processed image for identification afterward for different distress types. Table A.3 gives an overview of the pavement condition metrics which can be measured by video/image processing.

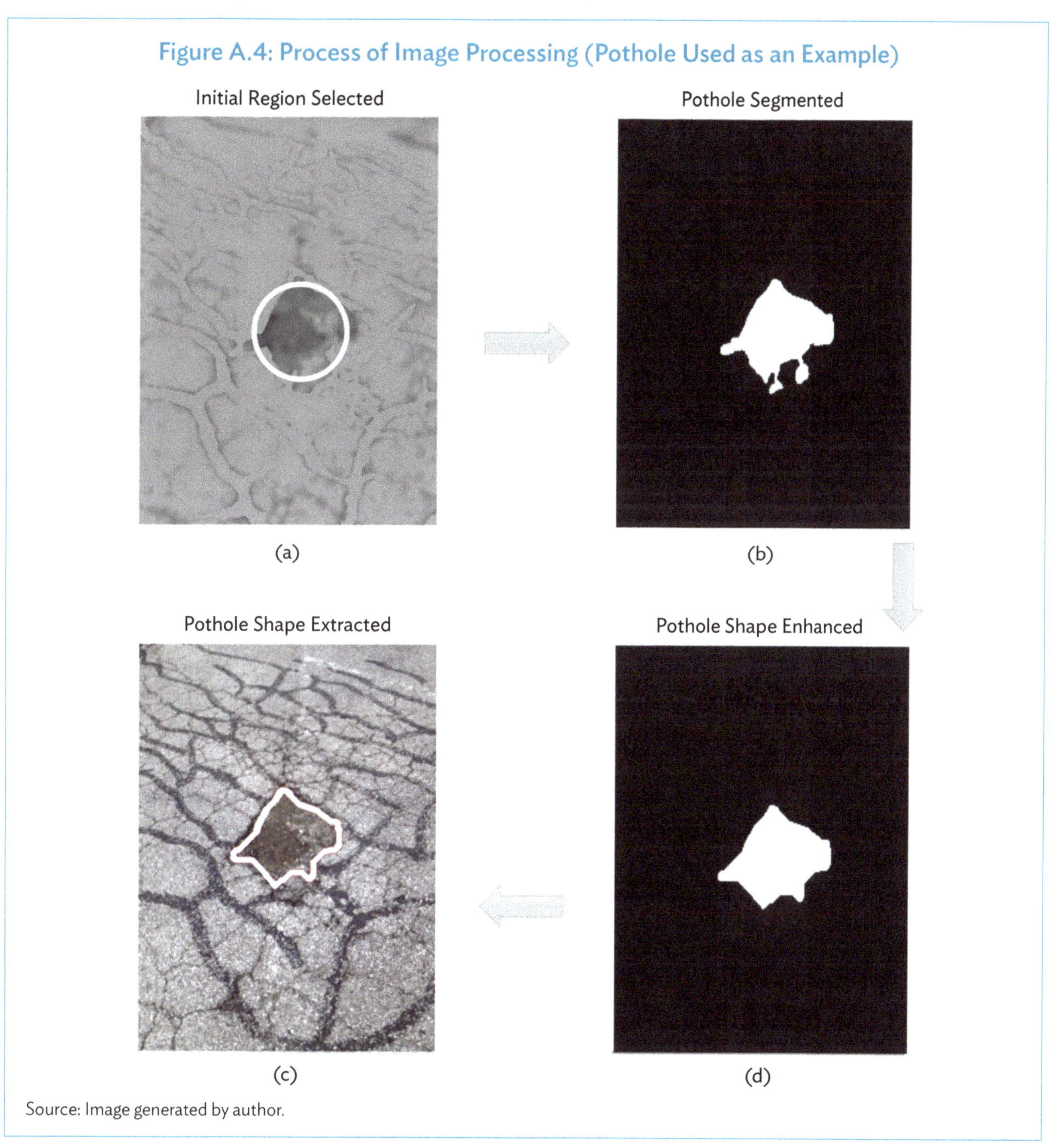

Figure A.4: Process of Image Processing (Pothole Used as an Example)

Source: Image generated by author.

Figure A.5: Raw Images Taken into the Analysis and the Processed Image for Identification for Different Distress Types (Cracking Used as an Example)

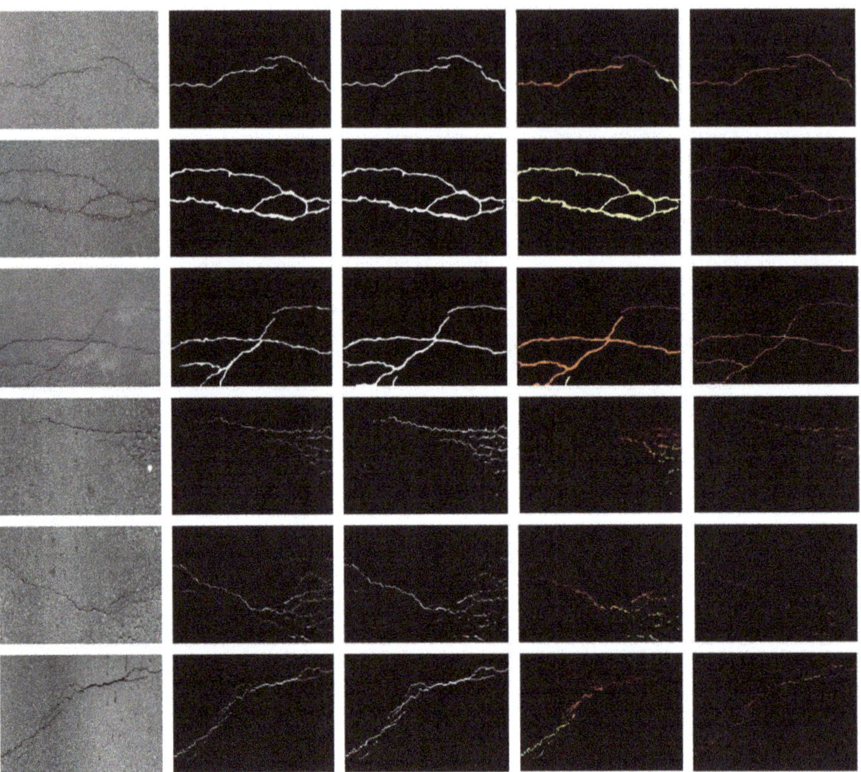

Source: Image generated by author.

Table A.3: Overview of Pavement Condition Metrics Measured Through Video and Image Processing

Pavement Condition Metrics Assessed	Attributes Measured for Each Metric	Case Study/Road Agency
Distress	Determine the existence of potential crack or distress features and their rough extent within the video image scenes	Northern Ontario region, Canada[a]
	Identify and categorize cracks and potholes into different severity levels	Baghdad[b]
	Automatically detect cracks in hot-mix asphalt (HMA) and Portland cement concrete (PCC) surfaced pavement	US and Canada[c]
	Automatically detect and characterize cracks in flexible pavement surfaces	Portugal[d]
	Automatically detect and characterize cracks potholes and patches	Bochum, Germany[e]
	Automatically detect potholes	Georgia[f]

[a] W. Yan and X. Yuan. 2018. A low-cost video-based pavement distress screening system for low-volume roads. *Journal of Intelligent Transportation Systems*, 376–389.

[b] I. Abbas and M. Ismael. 2021. Automated Pavement Distress Detection Using Image Processing Techniques. *Engineering, Technology & Applied Science Research*, 7702–7708.

[c] K. Gopalakrishnan et al. 2017. Deep Convolutional Neural Networks with transfer learning for computer vision-based data-driven pavement distress detection. *Construction and Building Materials*, 322–330.

[d] H. Oliveira and P. Correia. 2013. Automatic Road Crack Detection and Characterization. *IEEE Transactions On Intelligent Transportation Systems*, 155–168.

[e] K. Doycheva, C. Koch, and M. König. 2017. Enabled Pavement Distress Image Classification in Real Time. *Journal of Computing in Civil Engineering*.

[f] C. Koch, G. Jog, and I. Brilakis. 2013. Automated Pothole Distress Assessment Using Asphalt Pavement Video Data. *Journal of Computing in Civil Engineering*, 370–378.

Source: Authors.

In the study conducted by Zhu et al., an unmanned aerial vehicle (UAV) platform with a high-resolution camera is used to collect pavement images. The UAV flight parameters, such as flight altitude and flight speed, and camera parameters were tuned before the test. Moreover, a flight altitude of 30 meters was identified as the optimum altitude to avoid obstacles that may be encountered during flight over a road section, such as traffic signs, trees, and utility poles. Figure A.6 shows an example of a pavement image captured from a UAV. The study method has identified six pavement distress types, such as transverse crack, longitudinal crack, alligator crack, oblique crack, pothole, and repairs (which include sealed cracks and repaired patches).

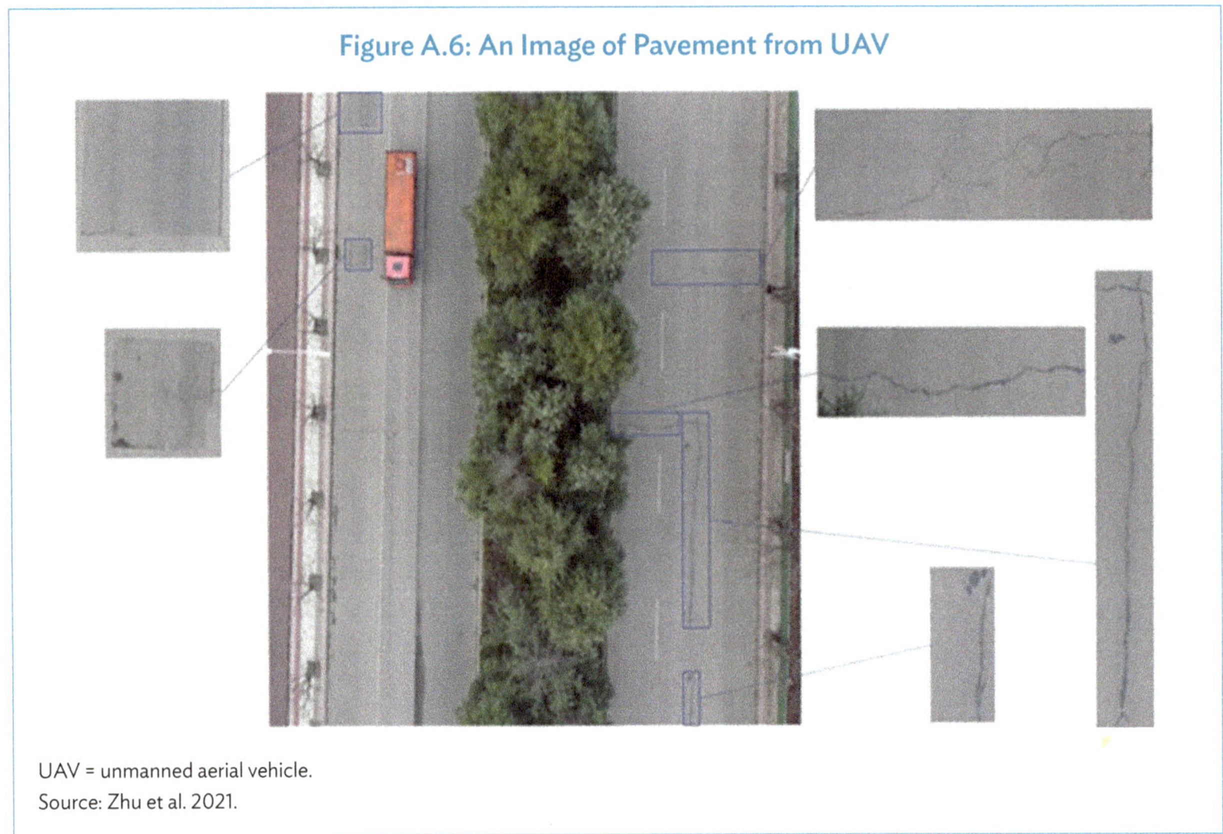

Figure A.6: An Image of Pavement from UAV

UAV = unmanned aerial vehicle.
Source: Zhu et al. 2021.

The objective of pavement distress detection is to classify and localize pavement distress from collected images, providing specific information for pavement maintenance and management. Deep-learning methods were used for pavement distress detection. Three typical object-detection algorithms were used: Faster R-CNN, YOLOv3, and YOLOv4.

Among the three algorithms, the prediction performance of YOLOv3 was the best, with an MAP of 56.62%. A typical pavement distress detection result using YOLOv3 model is shown in Figure A.7. The aerial images would be a good initiation, improving the image quality satellite images also can be used to evaluate pavement condition that would save significant amount of time and cost of surveys.

Figure A.7: A Typical Pavement Distress Detection Result Using YOLOv3 Model:
a) True-Positives of Crack, b) True-Positives of Repair, c) True-Positives of Pothole,
d) False-Positive Results, and e) Typical False-Negative Results

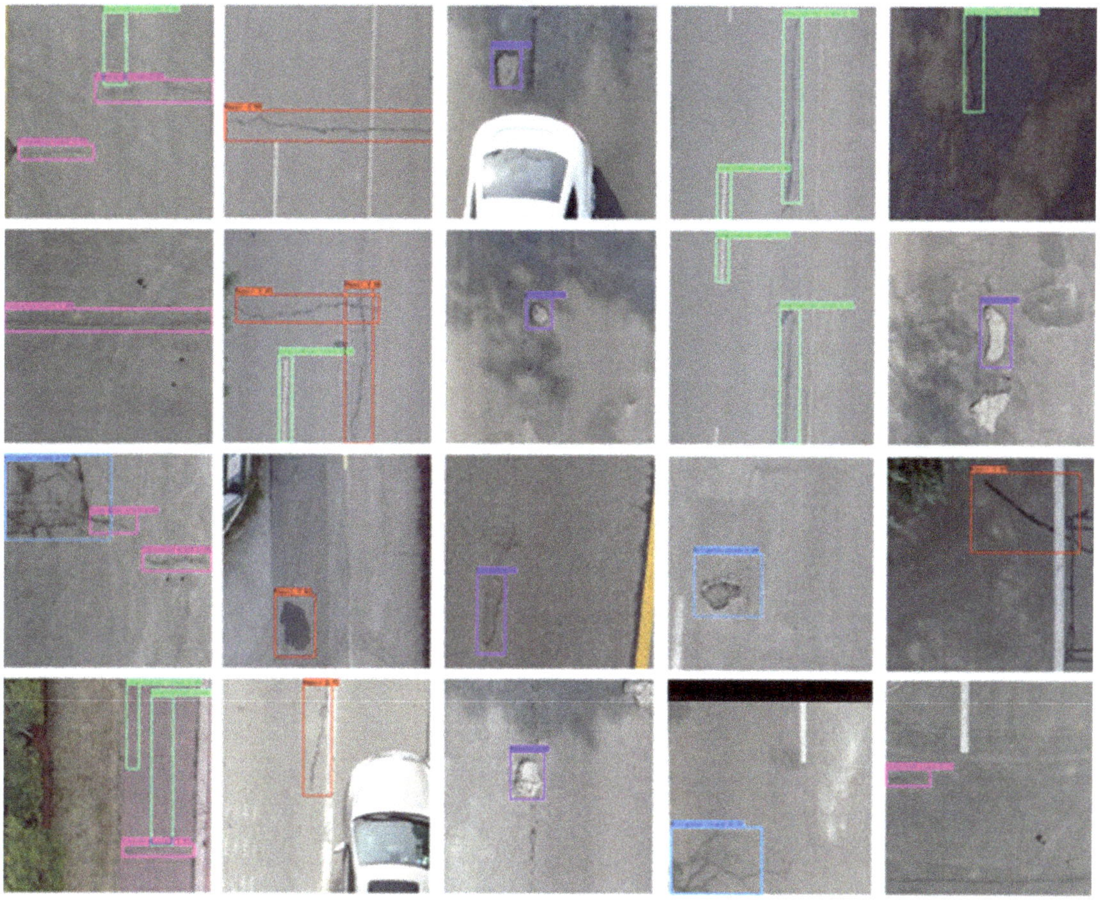

Source: Zhu et al. 2021.

Ground Penetrating Radar (GPR)

GPR is an electromagnetic-based, nondestructive, geophysical tool that is widely used for pavement investigations, including the differentiation of pavement layers, the mapping of layer thicknesses and the assessment of pavement condition.[16] GPR is primarily used to determine the pavement thickness and, subsurface defects in most situations, while some researchers have evaluated the effectiveness of GPR to measure the pavement surface distresses.

---

[16]   K. Aleksey et al. 2017. Utilization of air-launched ground penetrating radar (GPR) for pavement condition assessment. *Construction and Building Materials*, 130–139.

A typical GPR system has two major types of antennas: air-coupled and ground-coupled antennas.[17] Air-coupled antenna transmit high frequency signals, i.e., 2.0 gigahertz (GHz), while ground-coupled antennas can transmit small frequencies, i.e., 900 megahertz (MHz) and below. The high-frequency antenna can only scan the asphalt layer at shallow depth, and GPR generates high-frequency pulsed electromagnetic waves that penetrate the pavement structure as illustrated in Figure A.8.

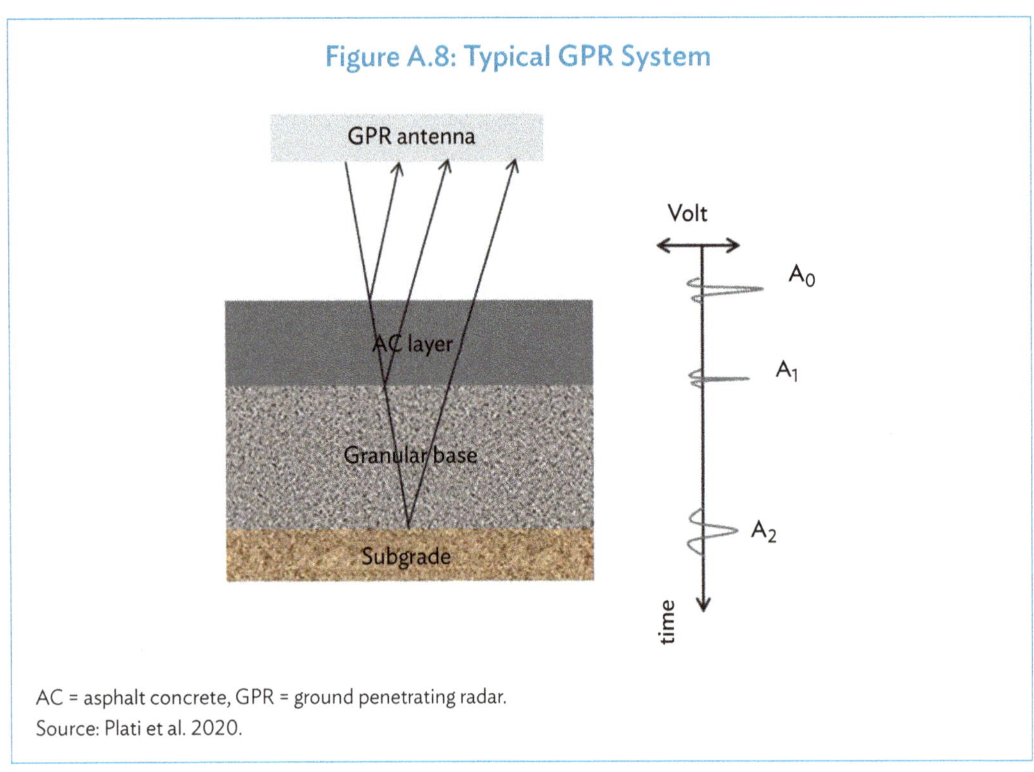

AC = asphalt concrete, GPR = ground penetrating radar.
Source: Plati et al. 2020.

GPR configuration and system usually measuring in a distance of 10 scan per meter (scan/m) is used for basic investigation, where the objective is to evaluate the subsurface. On the other hand, 20 scan/m is used in case of a more detailed investigation. In the situation where the GPR is used to detect pavement distresses, the measurement should be increased to 50–100 scan/m. As an example, Domitrović et al.[18] observed that at scan densities of 20–50 scan/m, potholes are already noticeable, and increasing the density to 100 scan/m gives a clearer insight into the extent of the distress.

---

[17]   T. Saarenketo and T. Scullion. 2000. Road evaluation with ground penetrating radar. *Journal of Applied Geophysics*, 119–138.

[18]   J. Domitrović et al. 2018. Investigating the use of GPR for pavement condition assessment. *5th International Conference on Road and Rail Infrastructure*, 495 - 501. Zadar: Pavements.

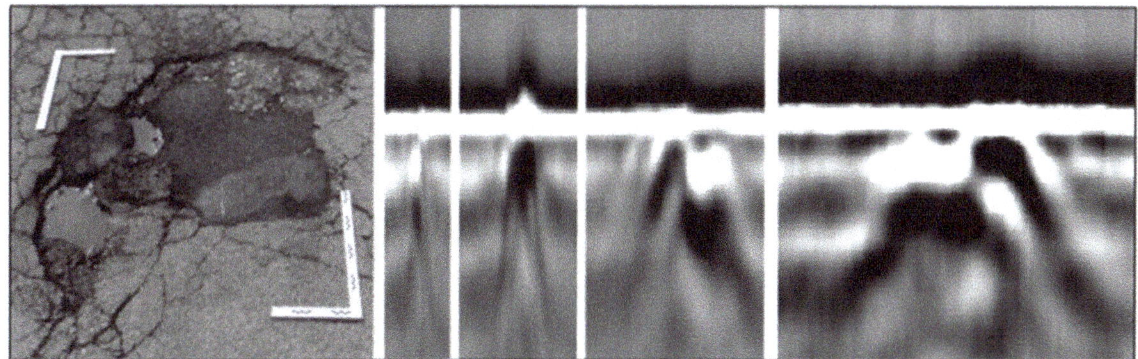

Figure A.9: Line Scan of Pothole Recorded with 10, 20, 50, and 100 Scan/M (from left to right)

Source: Domitrović et al. 2018.

Homogeneous zones of pavement surface can be identified by analyzing surface reflection amplitude and the amplitude of antenna movements with respect to the ground surface. Distress type identification is done by using a line scan and O-scope, and different patterns can be identified for different distress patterns as shown in Figure A.10. Table A.4 summarizes the overview of the pavement condition metrics which can be measured by GPR.

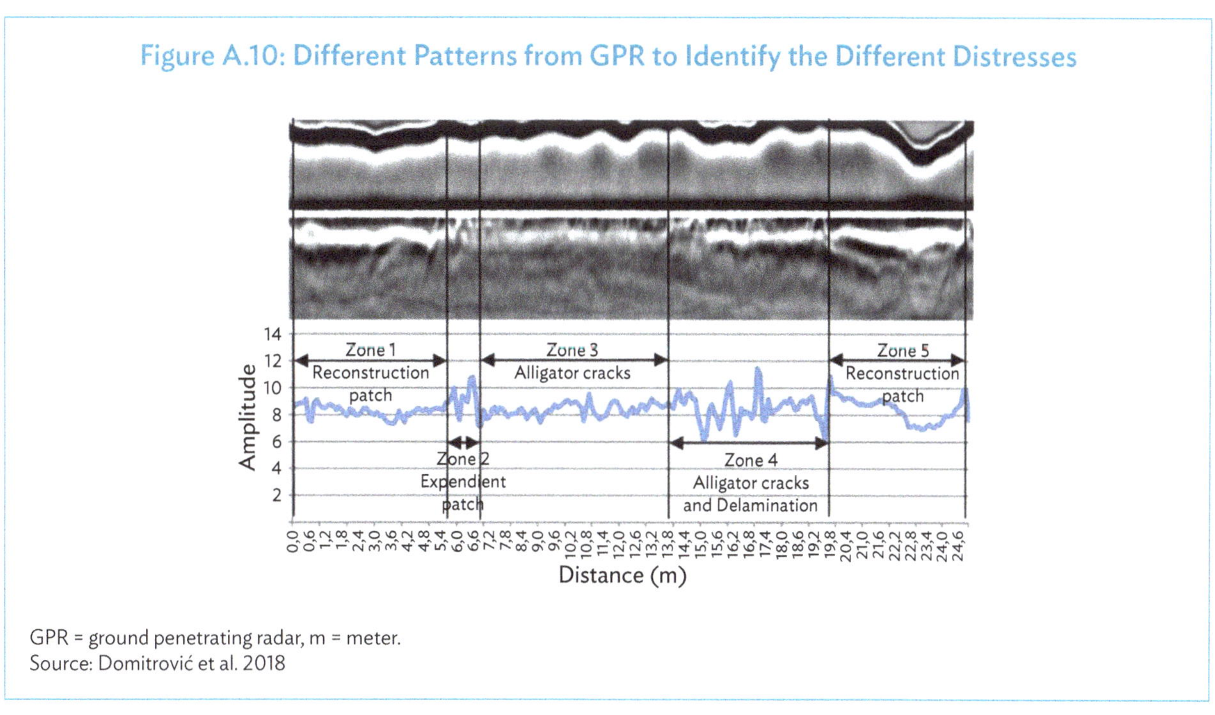

Figure A.10: Different Patterns from GPR to Identify the Different Distresses

GPR = ground penetrating radar, m = meter.
Source: Domitrović et al. 2018

Table A.4: Overview of Pavement Condition Metrics Measured Through GPR

Pavement Condition Metrics Assessed	Attributes Measured for Each Metric	Case Study/Road Agency Using This
Distress	Detect and measure cracks, water-damaged pits and uneven settlements.	Case study in the People's Republic of China[a]
	Detect cracks	Study conducted by Torbaghan et al.[b]
Pavement Profile	Estimate thickness and map the pavement material layers. Identify the presence of debonding, relative condition of the interface.	Case study in central Missouri[c]
	Pavement thickness Determine and identify underground objects	Minnesota Department of Transportation (MnDOT)[d]
	Determine density and thickness profile of asphalt concrete layer.	Case study in Macomb, Illinois[e]
	Determine layer thickness of flexible pavements.	Virginia Department of Transportation[f]

[a] Tong, Z., Yuan, D., Gao, J., Wei, Y., & Dou, H. 2019. Pavement-distress detection using ground-penetrating radar. *Construction and Building Materials*, 117352–117363.
[b] Torbaghan, M. E., Li, W., Metje, N., Michael, B., David, C. N., & Christopher, R. D. 2020. Automated detection of cracks in roads using ground penetrating radar. *Journal of Applied Geophysics*.
[c] Aleksey, K. K., Aleksandra, V. V., Evgeniy, T. V., Neil, A. L., & Lesley, S. H. 2017. Utilization of air-launched ground penetrating radar (GPR) for pavement condition assessment. *Construction and Building Materials*, 130–139.
[d] Dai, S., & Yan, Q. 2014. Pavement Evaluation Using Ground Penetrating Radar. *Pavement Materials, Structures, and Performance*, 222–231.
[e] Wang, S., Zhao, S., & Imad, A.-Q. L. 2018. Continuous Real-Time Monitoring of Flexible Pavement Layer Density and. *NDT and E International*.
[f] Al-Qadi, I., Lahouar, S., & Loulizi, A. 2005. *Groundpenetrating radar calibration at the Virginia Smart Road and signal analysis to improve prediction of flexible pavement layer thicknesses.* Virginia Department of Transportation.

Source: Authors.

Vehicle-Response-Based Measurement Method

Vehicle-response-based measurement methods are largely applicable to roughness detection. Class III roughness measuring equipment such as the bump integrator is being used to measure roughness along with the GPS coordinates of the road segment. Moreover, the vehicle responses can also be used for pothole detection and road profile classification. There are, however, several factors that can affect the outputs from vehicle response systems, such as vehicle sprung and un-sprung mass, suspension, tire pressure, etc.

Roughness is measured by using vehicle body vibration counting over a measured distance from odometers. Also, vertical acceleration component is a useful measurement parameter which can be used to estimate roughness. Potholes are detected by using acceleration components (in x, y, and z directions) and training that to identify potholes over other defects such as bumps, rutting, etc.

Table A.5 illustrates the pavement condition metrics that can be evaluated by using the vehicle response-based measurements.

Table A.5: Overview of Pavement Condition Metrics Measured Through Vehicle-Response-Based Measurement

Pavement Condition Metrics Assessed	Attributes Measured for Each Metric	Case Study/Road Agency
Roughness	PSD of vertical displacement, pulse count, RMS acceleration	McCann and Nguyen[a] United States[b] Ngwangwa, Heyns, Labuschagne, and Kululanga[c] People's Republic of China[d, e, f] Ireland[g] Japan[h]
Pothole	Z-peak (peak acceleration) Z- threshold (acceleration threshold)	Mehta, Mathur, Agarwal, Sharma, and Prakasha[i] Taipei,China[j] United States[k] Argentina[l], the People's Republic of China[m, n]

PSD = power spectral density, RMS = root mean square.

[a] R. McCann and S. Nguyen. 2007. System identification for a model-based. *IEEE region 5 technical conference*, 336–343.

[b] P. Sauerwein and B. Smith. 2011. Investigation of the implementation of a probe-vehicle based pavement roughness estimation system. *Charlottesville*.

[c] H. Ngwangwa et al. 2010. Reconstruction of road defects and road roughness classification using vehicle responses with artificial neural networks simulation. *Journal of Terramechanics*, 97–111.

[d] Z. Li, W. Yu, and X. Cui. 2018. Online classification of road roughness conditions with vehicle unsprung mass acceleration by sliding time window. *Shock and Vibration*

[e] Y. Du et al. 2014. Measurement of International Roughness Index by Using Z-Axis Accelerometers and GPS. *Mathematical Problems in Engineering*.

[f] J. Li, Z. Zhang, and W. Wang. 2019. New approach for estimating international roughness index based on the inverse pseudo excitation method. *Journal of Transportation Engineering Part B: Pavements*.

[g] A. González et al. 2008. The use of vehicle acceleration measurements to estimate road roughness. *Vehicle System Dynamics*, 483–499.

[h] N. Abulizi et al. 2016. Measuring and evaluating of road roughness conditions with a compact road profiler and ArcGIS. *Journal of Traffic and Transportation Engineering (English Edition)*, 398–411.

[i] J. Mehta et al. 2017. Pothole detection and analysis system (PODAS) for real time data using sensor networks. *Journal of Engineering and Applied Sciences*, 3090–3097.

[j] H. Wang et al. 2015. A real-time pothole detection approach for intelligent transportation system. *Mathematical Problems in Engineering*.

[k] J. Erikkson et al. 2008. The pothole patrol: using a mobile sensor network for road surface monitoring. *MobiSys'08*.

[l] A. Monteserin. 2018. (2018). Potholes vs. speed bumps: A multivariate time series classification approach. In M. Lykourentzou and HFTA, eds. CEUR workshop proceedings, 36–40. CEUR-WS.

[m] G. Xue et al. 2017. Pothole in the dark Perceiving pothole profiles with participatory urban vehicles. *IEEE Transactions on Mobile Computing*, 1408–1419.

[n] J. Ren and D. Liu. 2017. PADS: A reliable pothole detection system. *Lecture Notes in Computer Science (including subseries Lecture Notes in Artificial Intelligence and Lecture Notes in Bioinformatics)*, 327–338.

Source: authors.

Thermal Mapping

Thermal mapping is a vehicle-based survey technique to identify and quantify the distribution of temperature differences across a highway network. In this method, the spatial variation of road surface temperature is accurately measured using a high-resolution infrared thermometer, which it uses to identify patterns of temperature variation. Synchronized thermography are then developed to evaluate the pavement condition as well as the prevailing distress. Defects on the road pavement can be recorded with respect to different temperatures on those places. The asphalt pavement temperature changes due to ambient temperature changes and solar radiation during the day. Defects may cause changes in the heat transfer in the road pavement. In favorable conditions, these changes can be recorded by a thermal camera. The measurements can be performed for almost any time in the whole year under different weather conditions. However, the best results are achieved on sunny days, when the weather helps to create a temperature contrast on the road surface. Moreover, the survey speed has a direct impact on the efficiency of the data as well as the infrared resolution and frame rate.

The individual thermograph can be used to identify key distresses such as potholes and cracking on the pavement. Figure A.11 shows a thermograph of pavement with extensive potholes, while Figure A.12 shows the detection of cracks on the pavement.[19]

Figure A.11: Detection of Pothole with Thermal Mapping a) Thermogram of Pothole on Pavement and b) Digital Photo of the Same Pothole on Pavement

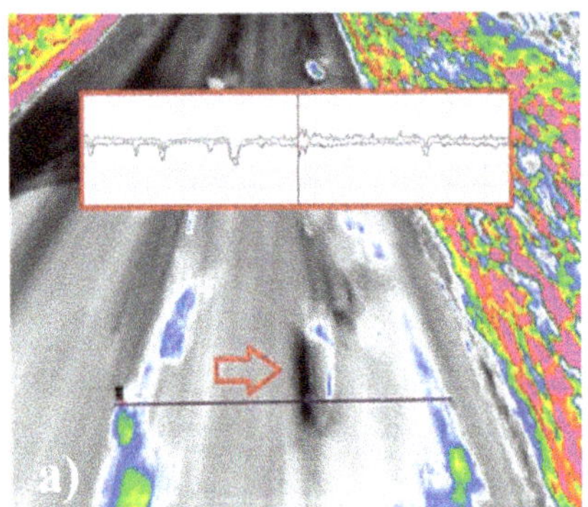

Source: Janků and Stryk, 2017.

---

[19]   Janků, M., & Stryk, J. 2017. Application of infrared camera to bituminous concrete pavements: measuring vehicle . *Materials Science and Engineering.*

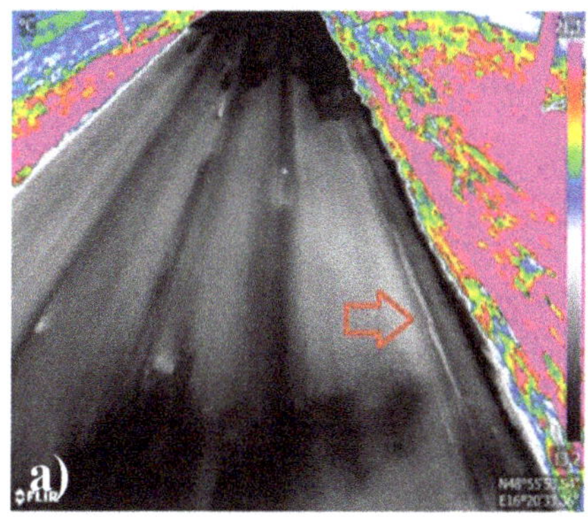

Figure A.12: Detection of Pothole with Thermal Mapping a) Thermogram of Crack on Pavement and b) Digital Photo of the Same Crack on Pavement

Source: Janků and Stryk, 2017.

Pavement Condition Assessment Techniques and Their Outputs

Table A.6: Comparison Between Automated and Manual Data Collection

Aspect	Manual Data Collection	Automated Data Collection
Time	Requires longer data collection time	Reduce data collection time
Safety	Personnel at risk when collecting data	Relatively safer data collection
Objectivity	Subjective measurement based on the experience of inspector	Objective measurement focusing on specific parameter
Cost	Less expensive	Expensive equipment cost both initial and operating cost
Data size	Only smaller amount of data can be collected at a time	Depend on the capacity of the equipment and large amount of data can be collected usually
Data handling	Subject to transcription errors	Not subject to transcription errors
Employers	Requires higher number of employees	Suitable in road agencies seeking to reduce the number of employees
Coverage	Inspection can cover the entire width at once	Can cover footprint of data collection, sometimes multiple runs needed to cover entire width

Source: authors.

The following Table A.7 summarizes the various condition assessment techniques, their operational requirements, key outputs and comparison of the resources to implement.

Table A.7: Important Characteristics of Recommended Equipment for Data Collection

Equipment	Principal of Operation	Output	Operating Speed	Multiple Measurement	Merits	Limitations	Recommendations
Benkelman Beam	Elastic deflection under static load	Rebound deflection at single point under load	Crawling	NA	Simple, quick, cheap	Single point deflection	Can be used for routine overlay requirements for all categories of roads
Falling Weight Deflectometer	Elastic deflection under impulse load	Deflection basin	200–300 stations per day	NA	Easy to operate relatively fast, complete deflection profile is measured	Expensive	Recommended for use on primary/ secondary road network
Loadman	Elastic deflection under impulse load	Single point deflection	80–100 stations per day	NA	Simple, portable	Single point deflection	Recommended on thin pavements with smooth surfaces subject to good correlations with already established methods
Automatic Road Analyzer (ARAN)	Continued images one forward and two straight down	Video tapes	30–100 kph	Distress (crack rut depth) profile, roughness, gradients, curvature, etc.	Covers about 600 km a day with good accuracy	Expensive, dry surface measurements Processing is manual	Recommended for use on primary/ secondary road network with dry surface
Laser Road Surface Tester (LRST)	Measures slope profile in the time domain	Distress unevenness international roughness index (IRI) long, profile	Up to 90 kph	Distress rut depth, profile, macro texture	Suitable for all weather condition	Calibration needed daily for laser and accelerometer	Recommended for all categories of roads
Towed fifth wheel bump integrator/ vehicle-mounted bump integrator	Response type measurement	Unevenness index	30 kph	NA	Simple, reliable data collection	Cannot measure profile needs frequent calibration	Recommended for all categories of roads

Equipment	Principal of Operation	Output	Operating Speed	Multiple Measurement	Merits	Limitations	Recommendations
Dipstick	Measures vertical profile by slope measurements	IRI	Slow walking	NA	Measures true profile	Very slow	Recommended for more accurate surface profile measurements and for calibration of response type equipment (towed/vehicle-mounted bump integrators)
British Pendulum Skid Tester	Measures lateral friction by swing action	Skid resistance	Very slow	NA	Simple, portable	Noting and recording manual spots measurements only	Recommended for all categories of roads
Mu Meter	Measures side force coefficient	Friction value	Up to 150 kph	NA	Requires little traffic control	Ineffective during winter, in sharp curves and in steep grades	Recommended for primary roads
Static Weigh Pads	Load measured through load cells/load bars	Static loads	Normal	NA	Any traffic condition	Vehicles need to be stopped and aligned	Recommended for medium and low traffic loads
Weigh-In-Motion	Piezo electric sensors and capacitor-type sensor	Weights of moving vehicles	Normal	Speed, vehicle type	Traffic is not interrupted during studies	Proper calibration checks are required	Recommended for primary/secondary roads with smooth surfaces, steel rimed tires should be avoided
Instrumented Car	Measurement of Upgrade downgrade and directional change of vehicle	Cumulative rise/fall curvature	32 kph	Unevenness, gradients, curves	Capturing of more parameters in single run	Low speed, sudden jerks adversely effect	Recommended for all categories of roads

km = kilometer, kph = kilometers per hour, NA = not applicable.

Source: Authors

References

Abdelaziz, N., Abd El-Hakim, R. T., El-Badawy, S. M., and Afify, H. A. 2020. International Roughness Index prediction model for flexible pavements. *International Journal of Pavement Engineering*, 88–99.

Abbas, I. H. and Ismael, M. Q. 2021. Automated Pavement Distress Detection Using Image Processing Techniques. *Engineering, Technology & Applied Science Research*, 7702–7708.

Abeygunawardhana, C., Sandamal, R., and Pasindu, H. 2020. Identification of the Impact on Road Roughness on Speed Patterns for Different Roadway Segments. *Moratuwa Research Conference (MERCon)*, 425–430. Moratuwa: IEEE.

Abeywardana, H.M.A., Abeywikrama, U.M.J., Amarasinghe, P.T., Kumarasinghe, R.P.D., Dilum Bandara, H.M.N., and Pasindu, H.R. 2018. iRoads – Smartphone-Based Road Condition Monitoring. *Final Year Project Report*.

Abulizi, N., Kawamura, A., Tomiyama, K., and Shun, F. 2016. Measuring and evaluating of road roughness conditions with a compact road profiler and ArcGIS. *Journal of Traffic and Transportation Engineering (English Edition)*, 398–411.

Abutaleb, A. S. 1989. Automatic Thresholding of Gray-Level Pictures Using Two-Dimensional Entropy. *Computer Vision, Graphics and Image Processing, 47*, 22–32.

Acosta, J. A. 1991. Implementation of the Video Image Processing Technique for Evaluating Pavement Surface Distress. *MS Thesis, Case Western Reserve University*.

Adolfo, A. J., Ludwig, F. J., and Mullen, R. L. 1992. Low-Cost Video Image Processing System for Evaluating Pavement Surface Distress. *Transportation Research Record*, 63–73.

Aleksey, K. K., Aleksandra, V. V., Evgeniy, T. V., Neil, A. L., and Lesley, S. H. 2017. Utilization of air-launched ground penetrating radar (GPR) for pavement condition assessment. *Construction and Building Materials*, 130–139.

Al-Qadi, I., Lahouar, S., and Loulizi, A. 2005. *Groundpenetrating radar calibration at the Virginia Smart Road and signal analysis to improve prediction of flexible pavement layer thicknesses.* Virginia Department of Transportation, 3.

American Association for the Advancement of Science. nd. *High-Resolution Satellite Imagery Ordering and Analysis Handbook.* https://www.aaas.org/resources/high-resolution-satellite-imagery-ordering-and-analysis-handbook.

Apollo Mapping. nd. *Worldwide-4 Satellite.* https://apollomapping.com/worldview-4-satellite-imagery.

Artis, M., Strazdins, G., Zviedris, R., Kanonirs, G., and Selavo, L. 2011. Real Time Pothole Detection Using Android Smartphones with Accelerometers. 1–6.

ASTM. 2012. *Standard test method for measurement vehicular response to traveled surface roughness.* ASTM E1082–90.

Attoh-Okine, N. and Adarkwa, O. 2013. *Pavement Condition Surveys –Overview of Current Practices.* Delaware Center for Transportation, University of Delaware.

Australian Road Research Board (ARRB). 2000. *ARRB System.* https://arrbsystems.com/fact-sheet/hawkeye-2000/

Bajic, M., Shahrzad, P. M., Skar, A., Pettinari, M., and Levenberg, E. 2021. Road Roughness Estimation Using Machine Learning. *IEEE.*

Belzowski, B., and Ekstrom, A. 2015. *Evaluating Roadway Surface Rating Technologies.* Michigan Department of Transportation.

Bisconsini, D. R., Nunez, J. Y., Nicoletti, R., and Junior, J. L. 2018. Pavement Roughness Evaluation with Smartphone. *International Journal of Science and Engineering Investigations, 7,* 43–52.

Blasiis, M. R., Benedetto, A. D., Fiani, M., and Garozzo, M. 2021. Assessing of the Road Pavement Roughness by Means of LiDAR Technology. *coatings.*

Burningham, S. and N. Stankevich. 2005. Why Road Maintenance is Important and How to Get It Done. *Transport Notes Series.* 4. World Bank. https://openknowledge.worldbank.org/handle/10986/11779.

Byrne, M., Parry, T., Isola, R., and Dawson, A. 2013. Identifying road defect information from smartphones. *Road and Transport Research,* 39–50.

Cadamuro, G., Muhebwa, A., and Taneja, J.. 2018. *Assigning a Grade: Accurate Measurement of Road Quality Using Satellite Imagery.* arXiv. https://doi.org/10.48550/arXiv.1812.01699.

Chandra, S. 2004. Effect of road roughness on capacity of two-lane roads. *Review of Transportation Engineering.*

Chang, K., Chang, J., and Liu, J. 2005. Detection of pavement distresses using 3D laser scanning technology. *2005 ASCE International Conference on Computing in Civil Engineering.*

Dai, S., and Yan, Q. 2014. Pavement Evaluation Using Ground Penetrating Radar. *Pavement Materials, Structures, and Performance,* 222–231.

Denis, E. 2014. Pavement Condition Monitoring with Crowdsourced Connected Vehicle Data. *Transportation Research Record Journal of the Transportation Research Board,* 1 - 36.

Domitrović, J., Bezina, S., Rukavina, T., and Stančerić, I. 2018. Investigating the use of GPR for pavement condition assessment. *5th International Conference on Road and Rail Infrastructure* (pp. 495–501). Zadar: Pavements.

Dotdash Meredith publishing family. 2021. *Investopedia* . Retrieved from Crowdsourcing: https://www.investopedia.com/terms/c/crowdsourcing.asp

Douangphachanh, V., and Oneyama, H. 2013. A study on the use of smartphones under realistic settings to estimate road roughness condition. *Journal of the Eastern Asia Society for Transportation Studies.*

Doycheva, K., Koch, C., and König, M. 2017. Enabled Pavement Distress Image Classification in Real Time. *Journal of Computing in Civil Engineering.*

Du, Y., Liu, C., Wu, D., and Jiang, S. 2014. Measurement of international roughness index by using Z-axis accelerometers and GPS. *Mathematical Problems in Engineering.*

Eisenbach, M. et al. 2019. *Enhancing the Quality of Visual Road Condition Assessment by Deep Learning.* 26th World Road Congress.

Eriksson, J., Girod, L., Hull, B., Newton, R., Madden, S., and Balakrishnan, H. 2008. The pothole patrol: Using a mobile sensor network for road surface monitoring. *Breckenridge: MobiSys'08 - proceedings of the 6th international conference on Mobile systems, applications, and services, 29–39.*

Famili, A. 2020. *Pavement Surface Evaluation Using Mobile Terrestrial LiDAR Scanning Systems.* Graduate School of Clemson University.

Fortunatus, M., Onyango, M., Fomunung, I., McLean, A., and Owino, J. 2018. Use of a Smart Phone based Application to Measure Roughness of Polyurethane Stabilized Concrete Pavement. *Civil Engineering Research Journal.*

Gao, Y. 2017. Automatic extraction of pavement markings on streets from point cloud data of mobile LiDAR. *Measurement Science and Technology.*

Gebreegziabher, B. A. 2021. *Mapping Road Pavement Quality from Optical Satellite Imagery using Machine Learning.* Faculty of Geo-Information Science and Earth Observation of the University of Twente.

Gong, H., Sun, Y., Shu, X., and Huang, B. 2018. Use of random forests regression for predicting iri of asphalt pavements. *Construction and Building Materials, 890–897.*

González, A., O'Brien, E. J., Li, Y. Y., and Cashell, K. 2008. The use of vehicle acceleration measurements to estimate road roughness. *Vehicle System, 483–499.*

Gopalakrishnan, K., Siddhartha, K. K., Choudhary, A., and Agrawal, A. 2017. Deep Convolutional Neural Networks with transfer learning for computer vision-based data-driven pavement distress detection. *Construction and Building Materials, 322–330.*

Grimmer Software. 2021. *RoadBump*. http://www.grimmersoftware.com/roadbump.html

He, K., Zhang, X., Ren, S., and Sun, J.. 2015. *Deep Residual Learning for Image Recognition*. arXiv. https://doi.org/10.48550/arXiv.1512.03385.

Hesami, R., and McManus, K. J. 2009. Signal processing approach to road roughness analysis and measurement. *TENCON 2009–2009, IEEE Region 10 Conf. Singapore: IEEE.*

Holgado, B., Riveiro, D., Gonzalez, A., and Arias, P. 2017. Automatic Inventory of Road Cross Sections from Mobile Laser Scanning System. *Computer-Aided Civil and Infrastructure Engineering, 32*(1), 3–17.

Islam, S., Buttlar, W. G., Aldunate, R. G., and Vavrik, W. R. 2014. Measurement of Pavement Roughness Using Android-Based Smartphone Application. *Transportation Research Record: Journal of the Transportation Research Board*, 30–38.

James, M. 1988. Paflem Recognition. John Wiley & Sons.

Janani, L., Sunitha, V., and Mathew, S. 2020. Influence of surface distresses on smartphone-based pavement roughness evaluation. *International Journal of Pavement Engineering*.

Janků, M. and Stryk, J. 2017. Application of infrared camera to bituminous concrete pavements: measuring vehicle . *Materials Science and Engineering*.

Jeong, J. H., Jo, H., and Ditzler, G. 2020. Convolutional neural networks for pavement roughness assessment using calibration-free vehicle dynamics. *Computer-aided Civil and Infrastructure Engineering*, 1209–1229.

Jones, H., and Forslof, L. 2014. Roadroid: Continuous road condition monitoring with smart phones. *Journal of Civil Engineering and Architecture*, 485–496.

Kionix. 2015. *AN012 accelerometer errors*. http://kionixfs.kionix.com/en/document/AN012%20 Accelerometer %20Errors.pdf.

Koch, C., Jog, G. M., and Brilakis, I. 2013. Automated Pothole Distress Assessment Using Asphalt Pavement Video Data. *Journal of Computing in Civil Engineering*, 370–378.

Koichi, Y. 2014. Collecting Pavement Big Data by using Smartphone. *1st IRF Asia Regional Congress Paper Submission Form*.

Koichi, Y. 2016. *BumpRecorder*. https://www.bumprecorder.com/document/161021_BumpRecorder_ workshop.pdf

Kos, A., Tomažič, S., and Umek, A. 2016. Evaluation of smartphone inertial sensor performance for cross-platform mobile applications. *Sensors*, 477–492.

Kumarage, S. 2018. *Use Of Crowdsourced Travel Time Data In Traffic Engineering Applications*. Moratuwa: University of Moratuwa.

Leduc, E. and G. Assaf. 2022. Road Visualization for Smart City: Solution Review with Road Quality Qualification. *Internet of Things. 12 (100305)*. https://doi.org/10.1016/j.iot.2020.100305.

Li, J., Zhang, Z., and Wang, W. 2019. New approach for estimating international roughness index based on the inverse pseudo excitation method. *Journal of Transportation Engineering Part B: Pavements.*

Li, M., Faghri, A., Ozden, A., and Yue, Y. 2017. Economic Feasibility Study for Pavement Monitoring Using Synthetic Aperture Radar-Based Satellite Remote Sensing; Cost–Benefit Analysis. *Transportation Research Record: Journal of the Transportation Research Board*, 1–11.

Li, X., Chen, R., and Chu, T. 2014. *A Crowdsourcing Solution for Road Surface Roughness Detection Using Smartphones.* Texas A&M University.

Li, Z., Cheng, C., Kwan, M. P., Tong, X., and Tian, S. 2019. Identifying Asphalt Pavement Distress Using UAV LiDAR Point Cloud Data and Random Forest Classification. *International Journal of Geo-Information, 8(39).*

Li, Z., Yu, W., and Cui, X. 2018. Online classification of road roughness conditions with vehicle unsprung mass acceleration by sliding time window. *Shock and Vibration.*

Loizos, A. and Plati, C. 2008. Evoluational Process of Pavement Roughness Evaluation Benifiting from Sensor Technology. *International Journal on Smart Sensing and Intelligent Systems*, 370–387.

Marcondes, J. A., Snyder, M. B., and Singh, P. 1992. Predicting vertical acceleration in vehicles through road roughness. *Journal of Transportation Engineering*, 33–49.

Masini, B., Bazzi, A., and Zanella, A. 2018. A Survey on the Roadmap to Mandate on Board Connectivity and Enable V2V-Based Vehicular Sensor Networks. *Sensors.*

McCann, R. and Nguyen, S. 2007. System identification for a model-based. *IEEE region 5 technical conference*, 336–343.

Mehta, J., Mathur, V., Agarwal, D., Sharma, A., and Prakasha, K. 2017. Pothole detection and analysis system (PODAS) for real time data using sensor networks. *Journal of Engineering and Applied Sciences*, 3090–3097.

Mohan, P., Padmanabhan, V. N., and Ramjee, R. 2008. Nericell: using mobile smartphones for rich monitoring of road and traffic conditions. *6th ACM conference on Embedded network sensor systems, ser. SenSys '08*, 357–358.

Monteserin, A. 2018. Potholes vs. speed bumps: A multivariate time series classification approach. *Lykourentzou, M. G. Armentano, & HFTA (Eds.), CEUR workshop proceedings*, 36–40. CEUR-WS.

Ngwangwa, H. K., Heyns, P. S., Labuschagne, F. J., and Kululanga, G. K. 2010. Reconstruction of road defects and road roughness classification using vehicle responses with artificial neural networks simulation. *Journal of Terramechanics*, 97–111.

Nitsche, P., Stütz, R., Kammer, M., and Maurer, P. 2012. Comparison of Machine Learning Methods for Evaluating Pavement Roughness Based on Vehicle Response. *Journal of Computing in Civil Engineering.*

NOS. 2021. *What is lidar?* Retrieved from National Oceanic and Atmospheric Administration, US Department of Commerce: https://oceanservice.noaa.gov/facts/lidar.html

NOS. 2021. *What is lidar?* Retrieved from National Oceanic and Atmospheric Administration, US Department of Commerce: https://oceanservice.noaa.gov/facts/lidar.html

Oliveira, H. and Correia, P. L. 2013. Automatic Road Crack Detection and Characterization. *IEEE Transactions on Intelligent Transportation Systems,* 155–168.

Oshri, B., Hu, A., Adelson, P., X, C., Dupas, P., and Weinstein, J. 2018. *Infrastructure Quality Assessment in Africa using Satellite Imagery and Deep Learning.* CoRR.

Pavemetrics. n.d. *Pavemetrics.* Retrieved from Pavemetrics® Laser Crack Measurement System (LCMS®-2): https://www.pavemetrics.com/applications/road-inspection/lcms2-en/lcms-2-roughness-iri/

PaveTesting® Ltd. 2011. *Profiling & Digital Imaging.* United Kingdom: PaveTesting® Ltd.

Plati, C., Loizos, A. and Gkyrtis, K. 2020. Assessment of Modern Roadways Using Non-destructive Geophysical Surveying Techniques. *Surv Geophys* 41, 395–430.

Ravi, R., Bullock, D., and Habib, A. 2020. Highway and Airport Runway Pavement Inspection Using Mobile LiDAR. *The International Archives of the Photogrammetry, Remote Sensing and Spatial Information Sciences,* 349–356.

Ren, J. and Liu, D. 2017. PADS: A reliable pothole detection system. *Lecture Notes in Computer Science (including subseries Lecture Notes in Artificial Intelligence and Lecture Notes in Bioinformatics),* 327–338.

Ritchie, S. G. 1990. Digital Imaging Concepts and Applications in Pavement Management. *Journal of Transportation Engineering, 116*(3), 287–298.

RoadBotics. 2021. *Visualize and analyze your roads and other infrastructure assets.* https://www.roadbotics.com/

RoadData. 2022. *RoadData - measuring highway and motorway markings for retro-reflectivity information you can trust.* https://roaddata.co.nz/?page=home

Roadroid. 2022. *Roadroid.* https://www.roadroid.com/

ROMDAS Data Collection Ltd. 2011. *ROMDAS System.* Retrieved from Laser Profilometer: https://romdas.com/romdas-laser-profiler.html

Saarenketo, T. and Scullion, T. 2000. Road evaluation with ground penetrating radar. *Journal of Applied Geophysics,* 119–138.

Sabottke, C. and B. Spieler. 2020. The Effect of Image Resolution on Deep Learning in Radiography. *Radiology: Artificial Intelligence. 2 (1).* https://doi.org/10.1148/ryai.2019190015.

Sandamal, R., and Pasindu, H. 2020. Applicability of Smartphone-based Roughness Data for Rural Road Pavement Condition Evaluation. *International Journal of Pavement Engineering.*

Sattar, S., Li, S., and Champmon, M. 2018. Road Surface Monitoring Using Smartphone Sensors: A Review. *Sensors.*

Sauerwein, P. M. and Smith, B. L. 2011. Investigation of the implementation of a probe-vehicle based pavement roughness estimation system. *Charlottesville.*

Sayers, M. W., & Karamihas, S. M. 1998. *The Little Book of Profiling.* Federal Highway Administration.

Sensor Event. 2012. *Android Developer Reference.* http://developer.android.com/reference/android/hardware/SensorEvent.html

Silva, C., Masini, B., Ferrari, G., and Thibault, I. 2017. A survey on infrastructure-based vehicular networks. *Mobile Transportation System.*

SoftTeco. (2021). *SoftTeco Projects.* https://softteco.com/projects

Sollazzo, G., Fwa, T., and Bosurgi, G. 2017. An ANN model to correlate roughness and structural performance in asphalt pavements. *Construction and Building Materials,* 684–693.

Sršen, M. 2002. *Automatic road analyzer- ARAN.*

Street Bump. 2020. *Where's Street Bump being used?* http://www.streetbump.org/about

Sun, L. 2001. *Developing spectrum-based models for international roughness index and present serviceability index.* Journal of Transportation Engineering, 463–470.

Tan, M., Pang, R., and Le, Q. *EfficientDet: Scalable and Efficient Object Detection.* Proceedings of the IEEE/CVF Conference on Computer Vision and Pattern Recognition (CVPR), 2020, pp. 10781–10790

Thube, D. T., Jain, S. S., and Parida, M. 2007. Development of PCI Based Composite Pavement Deterioration Curves for Low Volume Roads in India. *Highway Research Bulletin, Indian Roads Congress.*

Tong, Z., Yuan, D., Gao, J., Wei, Y., and Dou, H. 2019. Pavement-distress detection using ground-penetrating radar. *Construction and Building Materials,* 117352–117363.

Torbaghan, M. E., Li, W., Metje, N., Michael, B., David, C. N., and Christopher, R. D. 2020. Automated detection of cracks in roads using ground penetrating radar. *Journal of Applied Geophysics.*

TotalPave. 2022. *Collect IRI and PCI with Your Smartphone.* https://totalpave.com/

TotalPave. 2021. *Laser Profiler vs Smartphone – Precise vs Accurate IRI Data.* Retrieved from TotalPave: https://totalpave.com/blog/laser-profiler-vs-smartphone-precise-vs-accurate-iri-data/

Yeganeh, S. F., Mahmoudzadeh, A., Azizpour, M. A., and Golroo, A. 2019. *Validation of Smartphone-Based Pavement Roughness Measures*. Cornell University.

Tsai, Y., Ai, C., Wang, Z., and Pitts, E. 2013. Mobile crossslope. *Transportation Research Record: Journal of the*, 53–59.

Tsai, Y., Jiang, C., and Wang, Z. 2012. Pavement Crack Detection Using High-. *7th RILEM International Conference on Cracking in Pavements*, 169–178.

United Nations. 2015. Resoution 70/1 Transforming our world: 2030 Agenda for Sustainable Development adopted by the General Assembly on 25 September 2015. United Nations Official Document 2015.

UP42. 2023. *A definitive guide to buying and using satellite imagery*. https://up42.com/blog/tech/a-definitive-guide-to-buying-and-using-satellite-imagery.

VAISALA. 2021. *Thermal Mapping*. Retrieved from VAISALA: https://www.vaisala.com/en/product/1381

Varshney, N. 2023. *The Hallucination Problem of Large Language Models*. https://medium.com/@nvarshney97/the-hallucination-problem-of-large-language-models-5d7ab1b0f37f

Wang, G., Burrow, M., and Ghataora, G. 2020. Study of the Factors Affecting Road Roughness Measurement Using Smartphones. *Journal of Infrastructure Systems*.

Wang, H., Cai, Z., Luo, H., Wang, C. L., Yang, W., Ren, S.,and Li, J. 2012. Automatic road extraction from mobile laser scanning data. *Computer Vision in Remote Sensing*, 136–139. 2012 International Conference on IEEE.

Wang, H. W., Chen, C. H., Cheng, D. Y., Lin, C. H., and Lo, C. C. 2015. A real-time pothole detection approach for intelligent transportation system. *Mathematical Problems in Engineering*.

Wang, S., Zhao, S. and Imad, A.-Q. L. 2018. Continuous Real-Time Monitoring of Flexible Pavement Layer Density and. *NDT and E International*.

Wang, W. W. n.d. *BIG Revamping Road Condition and Safety Monitoring with Smartphones*. https://olc.worldbank.org/system/files/WBG_BD_CS_Road_Monitoring.pdf

William, B. G. and Islam, S. 2014. *Integration of Smart-Phone-Based Pavement Roughness Data Collection Tool with Asset Management System*. USDOT Region V Regional University Transportation Center.

Woodman, O. J. 2007. *An introduction to inertial navigation*. https://www.cl.cam.ac.uk/techreports/UCAM-CL-TR-696.pdf

Xue, G., Zhu, H., Hu, Z., Yu, J., Zhu, Y., and Luo, Y. 2017. Pothole in the dark Perceiving pothole profiles with participatory urban vehicles. *IEEE Transactions on Mobile Computing*, 1408–1419.

Yan, W. Y. and Yuan, X. 2018. A low-cost video-based pavement distress screening system for low-volume roads. *Journal of Intelligent Transportation Systems*, 376–389.

Yang, X., Wang, T., Ren, X., and Yu, W. 2017. Copula-based Multi-dimensional Crowdsourced Data Synthesis and Release with Local Privacy. *GLOBECOM 2017 - 2017 IEEE Global Communications Conference*. IEEE.

Zhang, Z., Sun, C., Bridgelall, R., and Sun, M. 2018. Application of a Machine Learning Method to Evaluate Road Roughness from Connected Vehicles. *Journal of Transportation Engineering, Part B: Pavements*.

Zhou, Q., Okte, E., and Al-Qadi, I. 2021. *Predicting Pavement Roughness Using Deep Learning Algorithms*. Washington D.C.: Transportation Research Record.

Zhu, J., Zhong, J., Ma, T., Huang, X., Zhang, W., and Zhou, Y. 2021. Pavement distress detection using convolutional neural networks with images captured via UAV. *Automation in Construction*.